Kenton Cast Iron Toys:

The Real Thing in Everything But Size

by Charles M. Jacobs

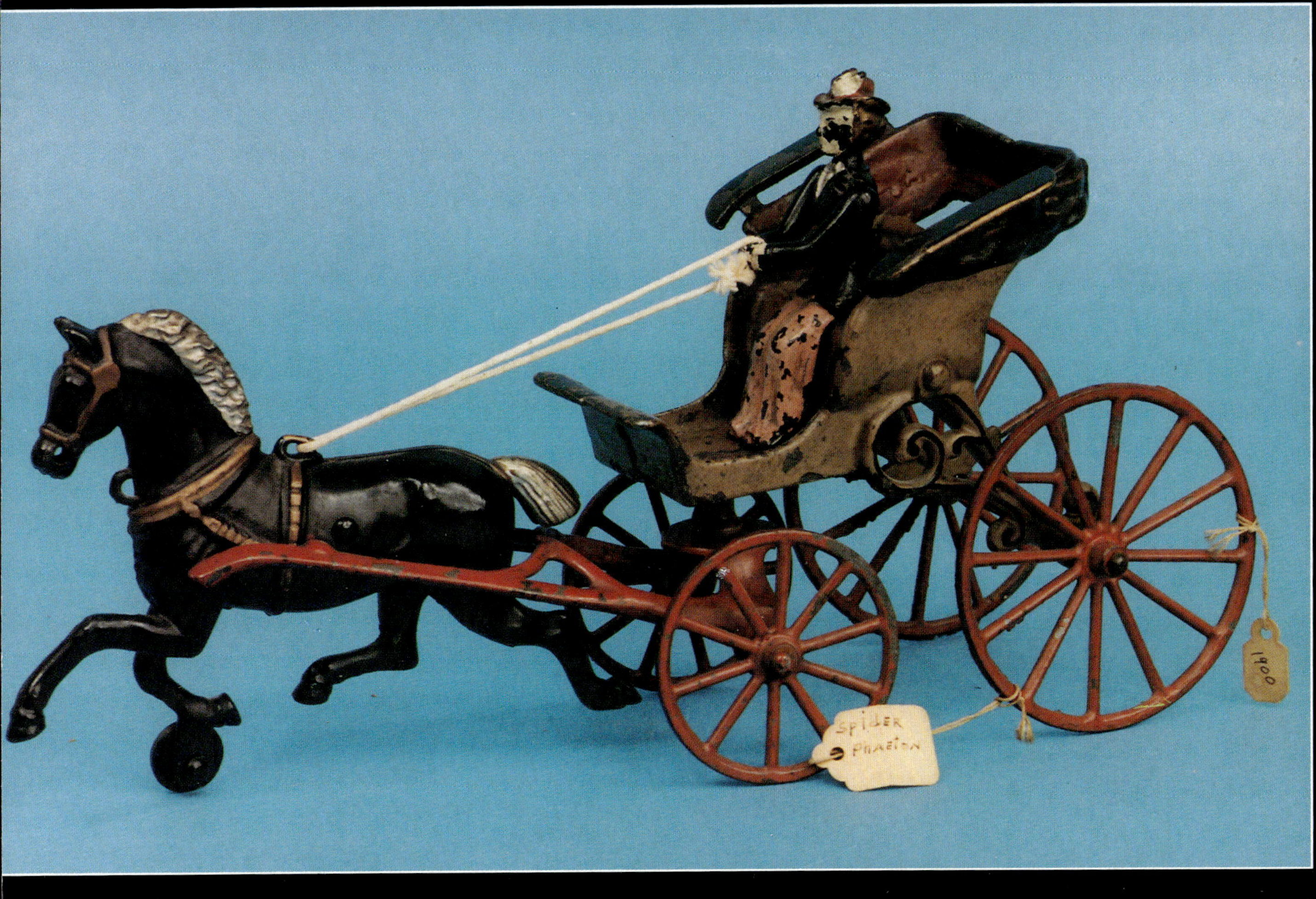

77 Lower Valley Road, Atglen, PA 19310

To my mother and father, who showed me that a red marble and a blue marble can open a whole world of imagination.

Printed in Hong Kong
ISBN: 0-88740-980-6

We are interested in hearing from authors
with book ideas on related topics.

Library of Congress Cataloging-in-Publication Data

Jacobs, Charles M. (Charles Matthew), 1956-
Kenton cast iron toys: the real thing in everything but size/by Charles M. Jacobs.
p. cm.
Includes bibliographical references and index.
ISBN 0-88740-980-6 (pbk.)
1. Kenton Hardware Company--History. 2. Metal toys--Ohio--Kenton--History. 3. Metal toys--Collectors and collecting--Catalogs. 4. Kenton Hardware Company. I. Title.
TS2301.T7J34 1996
688.7'2--dc20 96-1025
CIP

Published by Schiffer Publishing Ltd.
77 Lower Valley Road
Atglen, PA 19310
Please write for a free catalog.
This book may be purchased from the publisher.
Please include $2.95 postage.
Try your bookstore first.

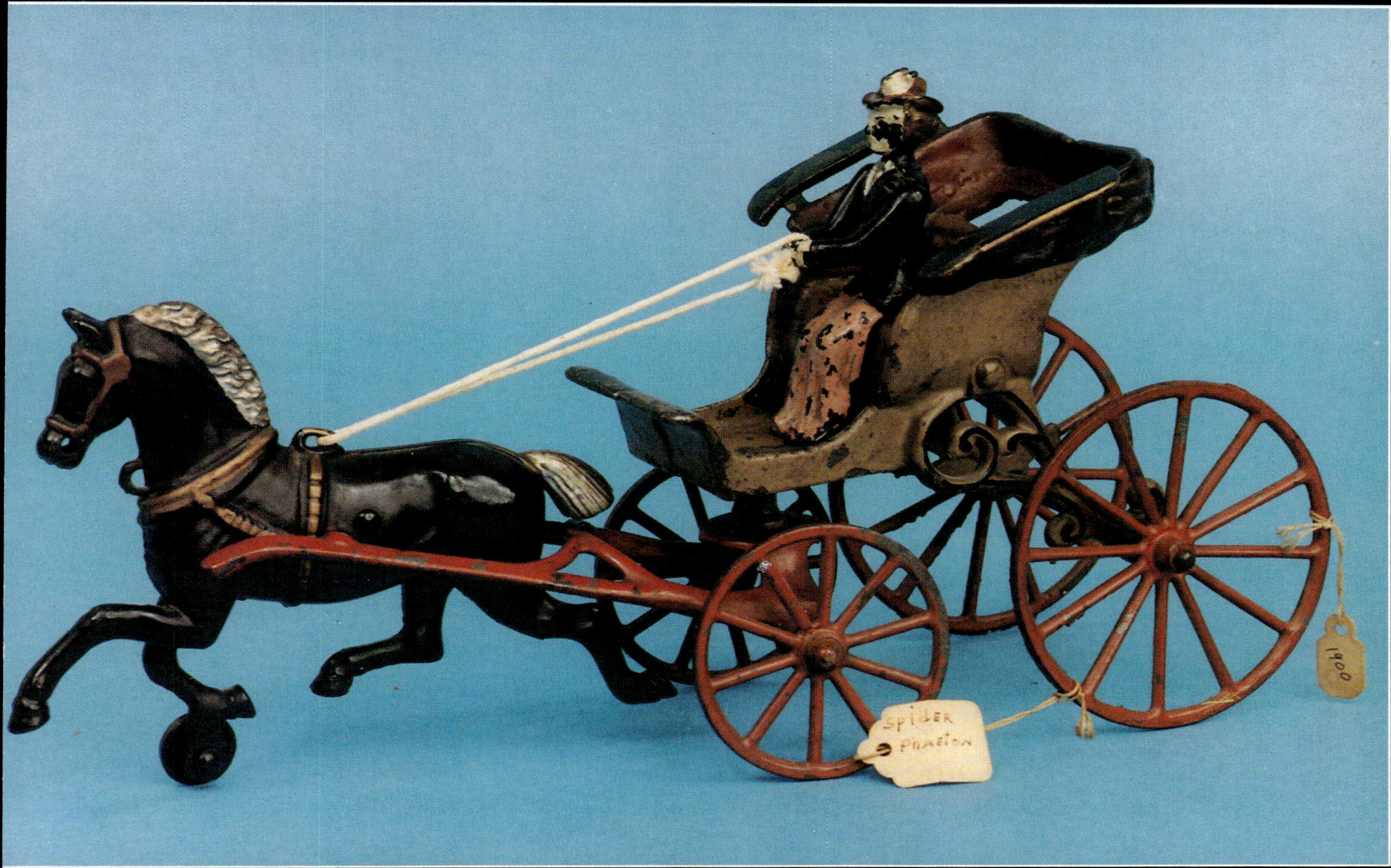

77 Lower Valley Road, Atglen, PA 19310

To my mother and father, who showed me that a red marble and a blue marble can open a whole world of imagination.

Printed in Hong Kong
ISBN: 0-88740-980-6

We are interested in hearing from authors
with book ideas on related topics.

Library of Congress Cataloging-in-Publication Data

Jacobs, Charles M. (Charles Matthew), 1956-
Kenton cast iron toys: the real thing in everything but size/by Charles M. Jacobs.
p. cm.
Includes bibliographical references and index.
ISBN 0-88740-980-6 (pbk.)
1. Kenton Hardware Company--History. 2. Metal toys--Ohio--Kenton--History. 3. Metal toys--Collectors and collecting--Catalogs. 4. Kenton Hardware Company. I. Title.
TS2301.T7J34 1996
688.7'2--dc20 96-1025
CIP

Published by Schiffer Publishing Ltd.
77 Lower Valley Road
Atglen, PA 19310
Please write for a free catalog.
This book may be purchased from the publisher.
Please include $2.95 postage.
Try your bookstore first.

Table of Contents

KENTON'S
NEW ENGRAVED
Gene Autry
REPEATER CAP PISTOLS
GENE AUTRY Sr.
Gene Autry
NO. 60
The cap pistol with the outstanding reputation. Just ask any youngster—they all know this model. It is an engraved scale model of Gene Autry's own six shooter. Nickel plated finish with brilliant red plastic grips. Gene Autry's script signature engraved on each grip. 9" long. Each pistol packed in an attractive three-color display box, showing a large photograph of Gene Autry. Packed 1/4 gross per shipping carton. Carton weight 37 lbs.
No. 60-C Gene Autry Sr. Same as No. 60 except in imitation gun metal finish with ivory plastic grips. Gene Autry's script signature engraved on each grip.
NO. 55
GENE AUTRY Jr.
Gene Autry
The Junior model of the regular Gene Autry pistol in a lower price range. Nickel plated finish with brilliant red plastic grips. Gene Autry's script signature engraved on each grip. This pistol has break action and appeal for every youngster. 7 1/4" long. Each packed in an attractive three color display box. Packed 1/2 gross per shipping carton. Carton weight 55 lbs.
No. 55-C Gene Autry Jr. Same as No. 55 except in imitation gun metal finish with ivory plastic grips. Gene Autry's script signature engraved on each grip.

Acknowledgements

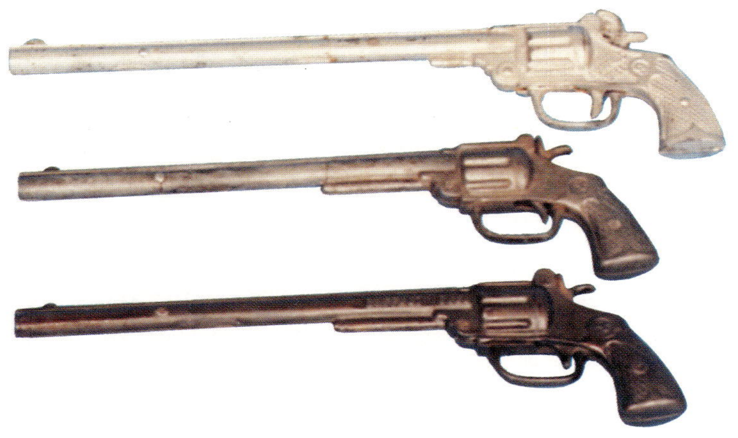

The most important contributors to this book are and were the hard-working men and women of Hardin County, Ohio, who conceived, manufactured, packaged, distributed, and promoted Kenton cast iron toys. For more than half a century, this dedicated guild brought prosperity and celebrity to their community.

I have been fortunate in having the assistance of faithful Kentontoy enthusiasts: my wife, Karen Craigo, who served as researcher, editor, typist, and booster, and wrote the profiles included as sidebars to chapters 9 and 10; the staff (Mary Sherman, Debra Boose, Jimmie Lightner, Charlie Pearson, and the late Dan Mathews), volunteers, and trustees of the Hardin County Historical Museums, Inc. of Kenton, Ohio, for their unflagging support; Gene Autry, for riding to the rescue once again; Bill and Jeannie Bertoia, for adding color and verve to a documentation of the early years; Dick and Joan Ford, for their hospitality and goodwill; Robert Saylor, for sharing his trailblazing research and three decades of collecting expertise; Tim Carrig of Golden Image Printing for his photographic wizardry; and Jeff Barnes of the *Kenton Times* for the use of his equipment.

In addition, I am extremely grateful to Don and Linda Houchin, J.E. Clark, Brad Bailey and Holly Sinn-Bailey and family, Dick Flowers, Glen Stearns, Charles Rogers, Jim and Teresa Geuy, Mark and Lynda Suozzi, Bob Bailey, Jack Herbert, Dan Allen, Eric Vasu-Sarver, Kenny Schlatter, Cliff Keel, Richard Etherton, Barney Bernard, Teresa Lay, and Bob and Terry Buroker, Jacqueline Murphy, Jack Herbert, and to John and Norma Sieg for filling many holes in my chronology. A special thanks to the staffs of the Mary Lou Johnson-Hardin County District Library, the Heterick Library of Ohio Northern University, the Detroit Public Library, the Hardin County Recorder's Office, Skinner, Inc., and TOYS etc.

This book would not have been possible without the Sally Wachalec and George Stevens bequest of Kenton Hardware Company archival materials and sample room collection, on permanent display at the Sullivan-Johnson Museum in Kenton, to the Hardin County Historical Museums, Inc.

Likenesses of Gene Autry and his licensed products are used with the permission of Mr. Autry. The Singing Cowboy, who saved a toy company, has helped me to preserve a rich and colorful heritage.

Preface

I have many special memories of the years I spent making personal appearances throughout the United States, Canada and overseas. One of the highlights was my trip to Kenton, Ohio in 1938.

I remember from the moment I arrived feeling right at home with the warm welcome I received. I was very impressed with the craftsmanship and care taken to produce the Gene Autry Cap Pistols. They were certainly a popular item with the children and I was glad to have the opportunity to visit the plant where they were manufactured.

Lewis Bixler and the hard-working employees of Kenton Hardware gave me quite a tour and an education. They left me in no doubt that things were in good hands and a fine agreement had been made by all of the parties concerned. I was doubly pleased to know that some difficult times had been eased for the company and a number of employees by the production of the Gene Autry Cap Pistol.

The heart and soul of a town are its people, and I will always treasure the friendships I made in Kenton that day. They were some of the best sidekicks a cowboy could ride with.

Gene Autry
July 11, 1995

GENE AUTRY 4383 Colfax Avenue. Studio City, California 91604

Introduction: What Makes a Kentontoy a Kentontoy?

From the 1890s to the 1950s, Kenton, Ohio was a center of American cast iron production, particularly in the field of toymaking. "Kentontoys," "KENTON TOYS," "KENTONTOYS," "KentonToys," "Kenton Cast Iron Toys," etc., were used interchangeably by the northwest Ohio toymaker, though the name of the line was never registered as a trademark.

A relatively small percentage of the playthings were marked. Early banks received preferential treatment (many bear an incised name of their maker), as did stoves, whereas wheel toys were for the most part left undesignated until the time of the Great Depression. Fortunately, a large percentage of Kenton catalogs have survived to aid identification. Seasoned collectors have familiarized themselves with the earmarks of the Kenton line: distinctive riveting, joinery, paint, modeling of figures, wheel and tire design, surface detailing, and high overall standards of quality. In addition, Kentontoys have an intangible quality that sets them apart from their competitors.

Kenton's use of catalog numbers must have been perplexing for its customers, and still presents a challenge for the researcher. In its formative years, the toymaker would graft new numbering systems onto the old. Arbitrarily, Kenton would change a product's number from one year to the next. The same number might be used for half a dozen different toys. For these reasons, the author has opted not to use the catalog numbers for identification.

On the other hand, Kenton was more consistent with its use of measurements. "OUR LINE herewith illustrated show actual lengths or sizes," began the 1915 catalog. The managers assured the company's clients, "We do not catalogue larger than actual dimensions, and on trains, the lengths given do not include coupling wires to add length to them." For the most part, Kenton's advertised dimensions have been found to be accurate, and the author has generally adopted them.

How many Kentontoys were manufactured? Various estimates have been made, ranging from three thousand to five thousand basic products. Recent discoveries (e.g., the jobbers' book) suggest that Kenton's output of penny toys and the like may have brought that figure closer to a staggering ten thousand. Orders for customized finishes would have boosted that total even more.

Chapter 1
Columbia, Gem of Kenton

Daniel Hudson Burnham and John W. Root designed the Administration Building as the hub and heart of Chicago's "White City," the centerpiece of the 1893 World's Columbian Exposition. The building's Civic Colossal style mingled Old World dignity and tradition with New World pomp and innovation. Its monumental proportions celebrated what President Grover Cleveland referred to as a "proud national destiny" and an "exalted mission" at the exposition's opening. The Administration Building, rising above an economic depression which cast a pall on the nation, heralded a new century and a new age of enlightenment for Chicago and all of the Midwest.

Kenton, Ohio, like Chicago, had experienced enormous growth since the Civil War. A prospectus published by a Kenton newspaper in the 1890s boasted of Kenton's upward growth in population and "steady and sure advancement in capitol [*sic*] and metropolis." Nestled between Toledo to the north and Columbus to the south and linked to these and other cities by a "complete network" of graveled pikes and rail lines, the Hardin County seat was "poised to become one of the leading trade centers [of northwest Ohio]."

Prospectus editor Daniel Flanagan and his colleagues observed, "Kenton's building blocks are principally of modern architecture, and in the great majority of cases are constructed of brick."

One such building block was the Kenton Lock Manufacturing Company in the southwest quadrant of the city. The concern had bucked the national economic distress and enjoyed success with its line of locks, latches, doorknobs, drawer pulls, and sash lifts. Its visionary captains of industry, in an optimistic, enterprising spirit befitting the World's Fair, looked to expand the product line. Their ambitions would lead to a name change for the factory and a change in focus. The *Columbia Bank*, a lilliputian version of the Administration Building, seemed a most fitting and proper choice as one of the first (if not the first overall) of Kenton's entries in the increasingly lucrative toy trade.

Toy Banks in America

New recreational opportunities and new wealth had inspired a golden age of toys in America during the late Victorian era. Pocket change expanded the market for toy banks, made first of tin plate and later of cast iron. Nonmechanical "still banks," with their pint-sized proportions and whimsical configurations, were appealing to youngsters; their instructive function–encouraging children to save for a rainy day–appealed to their parents, who had been brought up on lessons of Yankee thrift. When financial distress shook the nation in the early 1890s, the notions of economy and "squirreling" must have appealed to more and more Americans.

"Save your money" was one of the slogans for *The World's Fair Bank* marketed by Ives, Blakeslee, & Williams Company of Bridgeport, Connecticut, perhaps the toymaker nonpareil of the Gay Nineties. Bearing the likeness of a steam train and the legend "My Expenses to Chicago," this pottery globe was hailed as "a great novelty." The economic climate, coupled with the Columbian Exposition, made for a boom in toy banks. Kenton, too, jumped on the bandwagon.

The *Columbia Bank* may have been a novelty for Kenton, but it was hardly a unique contribution to the American toy market; miniature buildings provided the models for a large percentage of still banks produced in the late nineteenth century. Kenton found a crowded field of World's Fair banks: besides Ives, Blakeslee, & Williams, half a dozen or more companies produced a *Columbia* model. One concern molded a small glass Administration Building with a metal screw top. Another, the Magic Introduction Company, offered several varieties of a cast iron Administration Building with a detailed single facade.

However, Kenton's fine casting and extraordinary detail set its *Columbia* apart from the competition. Unlike the Magic Introduction line, Kenton's bank was intended to be seen in the round. In addition, the variety of finishes offered by Kenton evoked the splendor of the Administration Building at the World's Columbian Exposition as it appeared to fairgoers. A white paint treatment captured the look of the Great White City; aluminum and nickel-plated variations suggested the glimmer and shimmer of the palace under the night skies and above the artificial lighting which ringed the fair's canalway; and the electro-oxidized variations mimicked the inspiration's gilt dome and architectural elements. The overall verisimilitude embodied an age-old formula for success in the toy market: create the adult world in miniature, to give consumers toys that mirror daily life. The Civic Colossal architecture of the White City was to provide the model for thousands of buildings erected across the United States, and the Columbia Bank's styling mirrored urban tastes.

The Kentontoy Philosophy

Kenton was to echo this formula in its sales literature for much of its history. Four decades after the introduction of the *Columbia Bank*, Kenton promoters affirmed,

> The ideal toy should resemble some larger counterpart and, to some degree, instruct the child in its use. Also, it should be made of some material that is not easily destroyed. Kentontoys are popular because they meet both requirements.
>
> They are realistic. That is their principal appeal–they are copies of the things that are new which the child sees, reads, or hears about every day. Boys and girls are curiously interested in the things grown-ups are using in everyday life. With their modern ideas, they want Kentontoys, because they are up-to-the-minute and "look like the real thing."

Coupled with this formula was an emphasis on fine quality. The *Columbia Bank* and its successors, being of cast iron, were made to be longlasting and to "stand up under rough handling." Parents were assured that Kentontoys were built without sharp corners and edges, and were "perfectly safe for any child to enjoy."

Kenton realized that jobbers, the middlemen who bought from them and sold to other dealers, might associate Kenton quality with costliness. To counter this misapprehension, advertisers assured customers and potential customers that the *Columbia Bank* and its successors combined "the best construction" with a "reasonable cost."

Throughout much of its history, Kenton found cast iron to be a versatile, durable, and economical material. Made from iron alloyed with a small percentage of carbon and manganese (to increase hardness and to resist shock and embrittlement) and silicon (to add strength, rust-resistance, and hardness), cast iron lent itself to intricate shaping, painting, and plating. The relative ease of the casting process facilitated the mass production of component parts. For Kenton and its kindred toy manufacturers, cast iron was "good as gold" in the golden age of toys.

Columbia Bank: A Cornerstone of Kentontoys

Introduced in 1894 (or possibly in 1893), the *Columbia Bank* helped to propel Kenton into the American—and soon international–toy market. This miniature cast iron landmark, manufactured in a variety of finishes and sizes ranging from 4.5" to 9" in height, enjoyed a production which spanned a decade, approximately the same period of time that the Civic Colossal style remained in vogue. The *Columbia*, along with the Empire-designed *Home Savings Bank*, *New York National and Chicago Savings Bank*, and a handful of other combination banks, formed the vanguard for the Kentontoy line. If the Administration Building can be described as the hub of the World's Columbian Exposition, the *Columbia Bank* should qualify as a cornerstone of the Kentontoys industry.

Postal card depiction of the Administration Building.

Colonial Design.

No. 37. Aluminum, Height, $4\frac{1}{2}$ inches.
One in a box; three dozen in a case. Weight of case, **50 lbs.**

No. 38. Aluminum. Height 6 inches. **Combination lock on bottom.**
One in a box; three dozen in a case. Weight of **case, 100 lbs.**

No. 39. Aluminum. Height, $7\frac{1}{4}$ inches. **Combination lock on bottom.** One in a box; two dozen in a case. **Weight of case, 75 lbs.**

No. 40. Aluminum. Height, 9 inches. **Combination lock on bottom.**
One in a box; one dozen in a case. Weight of case, **75 lbs.**

Columbia Bank, as illustrated in an early twentieth-century catalog.

Left: *Columbia Bank* with a painted white finish. Right: a version which retains only traces of its nickel finish. Both measure 5.75" x 4.5" x 4.5".

Columbia Bank with aluminum finish, 5.75" x 4.5" x 4.5".

Centenarian Recalls Years of Change

The year was 1901. Guglielmo Marconi signaled the letter "S" across the Atlantic from England to Newfoundland. The revolutionary Emilio Aguinaldo was captured and the United States established civil government in the Philippines. President William McKinley attended the Pan-American Exposition in Buffalo, New York, and was struck down by an assassin's bullet.

It was also the year that Jessie Sheldon began working for the Kenton Hardware Manufacturing Company.

Eighty-six years later, when toy collector and historian Robert Saylor invited former Kenton Hardware employees to the opening of his exhibit "Kentontoys" at the Sullivan-Johnson Museum in Kenton, Ohio, he expected to meet workers with memories of the '30s, '40s, and early '50s. But he was also greeted by Jessie Sheldon Siniff, who was 100 years old on that day, March 28, 1987.

Jessie was just fourteen when she began work for the Kenton Hardware Manufacturing Company. Child labor laws had not yet been enacted. The early 1900s were a boom period for the factory. An extensive line of banks and toys were being cast, and the company began exporting to London, England.

Jessie was assigned to the packing room where finished playthings were sorted and boxed. Hours were long, and packing was difficult on both hands and feet. The job required standing on a hardwood floor.

During the seven years that she worked for the company at six dollars per week, Jessie witnessed many changes. A fire of suspicious origin raged through the plant in 1903. After rebuilding, changes in management prompted old toy lines to be dropped, including the first Kenton still bank, the *Columbia,* which Jessie remembered packing. Updated versions of toys were introduced. Jessie left the business just before it went bankrupt.

Saylor showed her a photograph of the packing room taken in circa 1905, which shows her with other girls before a long table topped with toy autos, fire equipment, and horse-drawn wagons. Mrs. Siniff produced another photograph, which she then donated to the museum, showing the girls at leisure, reclining in the yard adjacent to the factory.

Miss Siniff (left, with white apron) in packing room, early 1900s. *Courtesy of Hardin County Historical Museums, Inc.*

Jessie Sheldon Siniff with a large version of the *Columbia Bank,* 8.75" x 7" x 7". *Courtesy of the Kenton Times.*

Chapter 2
"It's a Go!"

In April of 1890, Floyd N. Perkins (for many years connected with lockworks in Cleveland and Chicago) met with Kenton's Board of Trade and other city leaders and proposed locating a large builders' hardware factory which would employ over two hundred people. The Upper Sandusky *Daily Union* on April 19 reported that Perkins required Kentonites to raise a bonus of $60,000 or more as an incentive for the company to locate there. A correspondent reported to his readers that the response of the citizens was "stranger than fiction." Being people "on the hustle" in a town "big as Barnum's greatest show on earth–in a fever of excitement–everybody on the go," Kenton was not flattered by the extravagant request. But the Board of Trade was determined to have the lock factory; so a few of the prominent citizens bought the old General James S. Robinson Farm near the Hardin County Fairgrounds in southwest Kenton, had it surveyed and platted, and sold lots for $200 each.

The *Kenton Daily News* editor implored his readers to support the effort.

> Everybody ought to take an active interest in this enterprise. It will benefit you, the city, and your posterity. It will be the means of bringing a number of intelligent people into our midst, and will cause house after house to be built here for dwelling purpose until the city of Kenton shall be knowst [*sic*] throughout the length and breadth of the land as one of the principal business points in Ohio. Citizens of Kenton, make up your minds to lend this factory your assistance. Buy one or more lots; it will be a safe investment.

The lock factory became a *cause celebre*, and was envisioned as the cornerstone of the new Riverside Addition, an industrial and housing development enhanced by a scenic park. The factory, together with the Riverside Addition, were heralded as the impetus for a great "booming" to the city.

In mid-May, a *Kenton Daily News* headline crowed, "It's a Go!" The article continued, "The Location of the Perkins Lock Factory in our City is Definitely Settled. . . . Col. J.C. Howe was sent to Cleveland this afternoon with full authority to make necessary arrangements and contracts for the purchase of machinery, and to have the same shipped at once. Operations will begin as soon as possible . . ."

The Perkins company lost little time in getting the business up and running. The newly-named Kenton Lock Manufacturing Company leased the former Forbings Scroll Mill on South Leighton Street while ground was being broken on the Riverside Addition project. A contemporary described the temporary quarters:

> The main room of the building is 60 by 25 feet and will be used as the finishing room and now is being furnished with machinery for that purpose. In the upstairs rooms the pattern-makers are already at work on the patterns for a special order, from the Baldwin Manufacturing Company of Boston, New York, Chicago, and Los Angeles. They are manufacturers of refrigerators and placed a trial order for 2,500 locks. The walls are completely covered with patterns of staple goods, locks, knobs, door plates, etc., and they probably number around 5,000.

Meanwhile, at the corner of Factory Avenue and Levings Street, work progressed rapidly on the new factory. As early as December of 1890, the local press described the Kenton Lock Manufacturing Company, a joint stock company with a capital stock of $75,000, as a prosperous and thriving establishment. Reporters were impressed by the larger brick facilities: a main building, 50' by 120'; an engine room, 35' by 40'; and a foundry, 45' square, as well as the "very latest" machinery with "all the modern improvements." They noted approximately fifty workers turning out locks, builders' hardware, and novelties in iron, brass, and bronze. They made special note of the factory's specialty, Munroe's patent refrigerator locks, "The best in the world, for which they are having a splendid sale at present."

In an address to a crowd of potential investors gathered at the Riverside Addition, the Honorable Frank C. Dougherty praised superintendent and manager Floyd N. Perkins and his brother and foreman, John, and warned, "Woe be unto the man who is a drawback to (their) enterprise. Better had a millstone be tied around his neck and that he be thrown into the Scioto (River)." The Perkinses, in turn, acknowledged their indebtedness to the skilled craftsmen who had worked for them in Cleveland or were recruited from as far away as New Britain, Canada.

The Perkinses and their colleagues entered a trade which had witnessed dramatic growth in the previous half century. The opening of natural gas, oil, coal, and ore fields in the Great Lakes states and their neighbors, as well as continuous technological advancements, provided the catalysts for a westward expansion of the cast metals industry. Despite periodic economic recessions and depressions, foundries sprouted up exponentially across the Midwest. There were linked by ever-improving railroads and waterways which brought in ever-depreciating raw materials and sent out an ever-broadening variety of metal wares. Contemporaries likened the metallurgical boom to a modern bronze or iron age.

The First Catalog

In 1892, the Kenton Lock Manufacturing Company introduced its first catalog to the trade. In a greeting to "friends and patrons," the officers boasted,

> . . . our business has increased five-fold in the last two years, necessitating three new additions to our factory (one of which is now in the process of construction). . . . We have every convenience for turning out first-class work, in entirely new designs, with our new and improved labor-saving machinery, having many special tools for cheapening work. Our natural advantages for manufacturing cheaply are many, being located in the great natural gas and oil belt, and on three trunk line railroads. We are the only Lock manufacturing concern west of the Allegheny Mountains that make a full line of Builders' Hardware; and our freight rates are much lower than our Eastern competitors.

A surviving example of the catalog provides proof that the Lock Manufacturing Company did live up to its claim of making a full line of builders' hardware. Among the 120-plus pages are pictured a bewildering array of brass, bronze, iron, and steel products: vestibule latches, slide bolts, mortise knob locks, door sets, escutcheons, key plates, pin butts, door knobs, push and pull plates, bar handles, window stops, lifts, pull sockets, coat and hat hooks, price card holders, letterbox plates, drawer and screen door pulls, shutter bars, cupboard and transom catches, coat and hat hooks, house numbers, ornamental table and chair feet, towel brackets, keys, glass or flower brackets, drawer knobs, name plates, and refrigerator locks, hinges, butts, and trimmings.

The trimmings category merited a supplemental catalog; the Kenton Lock Manufacturing Company noted it made a specialty of patents which closed and tightened icebox doors automatically, "making a perfect air-tight joint, something that no other Refrigerator Lock will do." The Alaska Refrigerator Company, Kankakee Manufacturing Company, Belding Company of Michigan, and the Baldwin Company (Burlington, Vermont) were among the satisfied clients that were cited. Kenton took special pride in "styles of finish," which included Light Bronze, Dark Background, and Chocolate Background, Polished Surface, Old Copper, Oxidized Silver, Antique Brass, the Natural Color of the Metal, Berlin Bronze, and a "very popular finish," Bronze Plated.

The Growth of the New Company

The *Kenton Daily News* reported in January of 1892 that more than eighty hands were employed by the Kenton lock factory, and that the managers expected to increase that number to a hundred by spring. In addition, the manager intended to branch out into plaster of paris and porcelain door knob production, necessitating an expansion of the facilities. "Immense amounts of raw materials" were consumed in the manufacture of locks and goods, which were shipped to "all parts of the country from the Atlantic to the Pacific oceans" (and even to Honolulu in the Sandwich Islands, according to an earlier item). The company was in good financial health with "book orders for goods aggregating $50,000." The weekly payroll amounted to $600. The article concluded, "The Lock Factory begins the new year under very favorable circumstances and the outlook is very bright for its success."

However, there was a cloud on the horizon: a nationwide depression which throttled most manufacturers. The lock factory's three-week shutdown in August 1893 suggests that Kenton was not spared (or did it perhaps indicate a retooling for a major overhaul in the product line?).

The Perkinses fell into a dispute with rivals and even with their own company officials over improvements to Munroe's patent lock. They found themselves mired in a lawsuit over license fees, and eventually left Kenton for Chicago. Nicholas H. Colwell was successor to superintendent-manager Floyd Perkins. Colwell had the managerial skills which his new position demanded. A graduate of Northwestern Ohio Normal School in Ada (near Kenton), he had served as civil engineer for the Mexican National Railroad. He had then returned to Ohio, became Hardin County's surveyor, and established a private engineering business.

An intriguing newspaper item from a Kenton newspaper of June 22, 1893, indicated that N.H. Colwell and family had returned from a trip to the Columbian Exposition in Chicago. One cannot help but try to connect that visit with the contemporaneous introduction of the *Columbia Bank*. Colwell would return to Chicago "in the interest of the company" the following spring, securing an order for two carloads of locks . . . and perhaps a quantity of banks.

A New Direction

With new management came new direction for Kenton. Following the lead of builders' hardware manufacturers A.C. Williams (Ravenna, Ohio) and the J. & E. Stevens Company (Cromwell, Connecticut), Colwell looked to diversify foundry work and capitalize on the growing demand for cast iron toys. Before 1894 was out, the Kenton Lock Manufacturing Company was rechristened the Kenton Hardware Manufacturing Company. Explained company officials, "Some of [our] trade have thought by the [former] name that only locks were manufactured." Writing nearly a decade after the name change, a Kenton journalist linked the change to a reorganization and the earnest expansion into "iron toys of all description." Locks were first eclipsed by a broader range of builders' hardware, which in turn was eclipsed, in just a matter of a few years, by cast iron toys.

The company continued to produce hardware well into the next century. Not a little incongruously, illustrations of a "practical and serviceable line of hat and coat hooks" appeared among those of wheel toys, ranges, floor trains, and banks in Kenton catalogs in the decade following 1910. Also not a little incongruously, the toy company kept its Kenton Hardware name. And as late as the 1920s, Kenton continued to advertise hardware products while maintaining that it was the largest factory in North America devoted to the exclusive manufacture of iron toys.

Locks had been the key to Kenton's early success in the national market. Cast iron toys, however, were to be the main force in Kenton's maturation as a big business and as a leader in world trade.

Kenton Lock Manufacturing Company plant as depicted in 1892 catalog.

Opposite page:

Bird's eye view of Kenton, 1891. Riverside Addition at left center features the newly-built Kenton Lock Manufacturing Company at the corner of Factory Avenue and Levings Street. The photo is inscribed "MacBriar Lith. Co., Cincinnati, O." *Courtesy of Hardin County Historical Museums, Inc.*

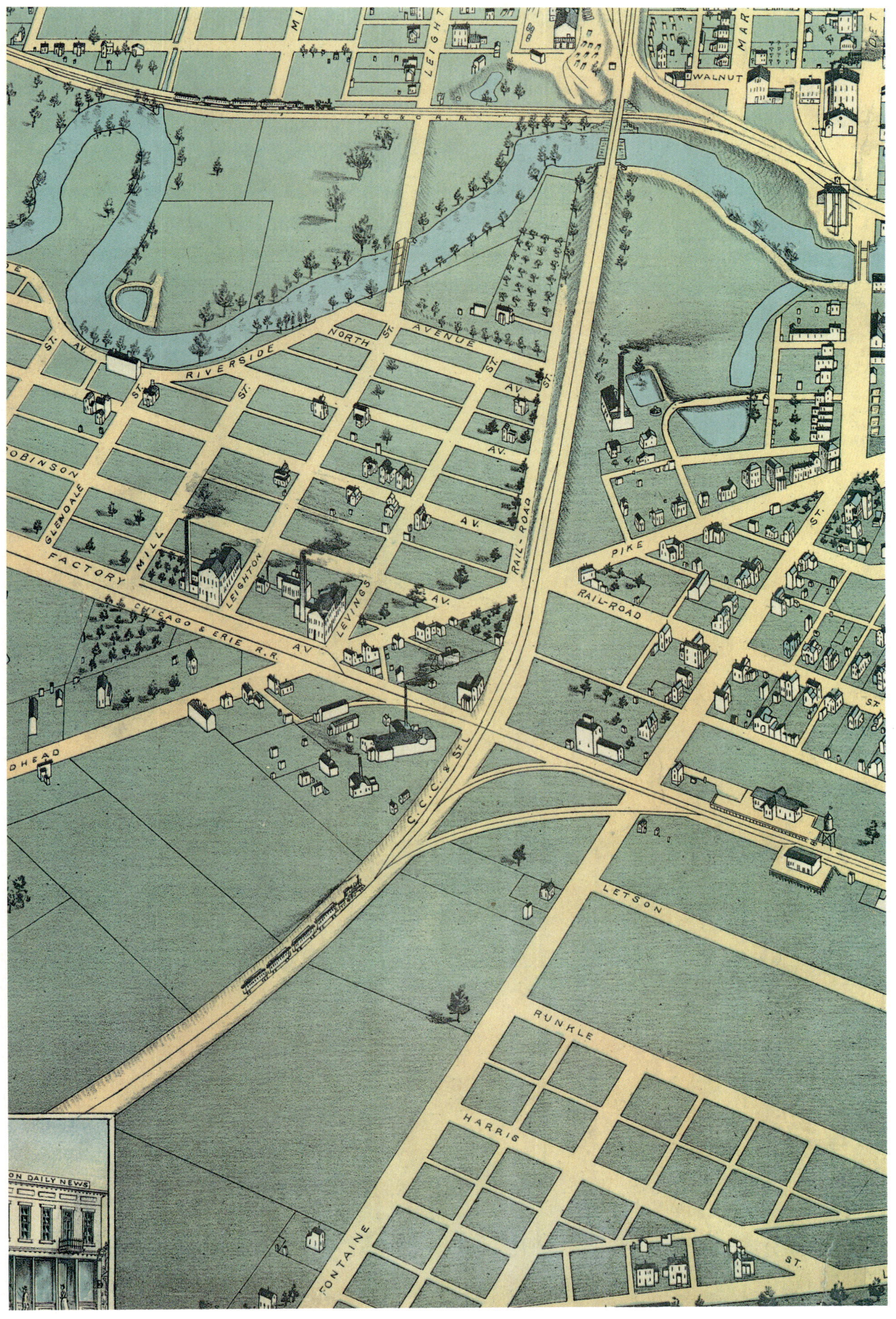
WALNUT
T. C. & C. R. R.
NORTH ST.
AVENUE
RIVERSIDE
ST.
AV.
ROBINSON
GLENDALE
FACTORY
MILL
LEIGHTON
LEVINGS
CHICAGO & ERIE R.R.
AV
RAIL-ROAD
PIKE
ST.
RAIL-ROAD
C.C.C. & ST. L.
LETSON
RUNKLE
HARRIS
FONTAINE
DAILY NEWS

Real Bronze or Brass Refrigerator Lock.

The Perkins' Wedging, Made Right or Left Handed.

Patented, November 24, 1891.

No. 9-0. LEFT HAND.

FINISHED IN ANY FINISH.

DESIGN
MADE FOR THE EXCLUSIVE USE OF THE ALASKA REFRIGERATOR CO.

No 8-0. RIGHT HAND.

FINISHED IN ANY FINISH.

DESIGN
MADE FOR THE EXCLUSIVE USE OF THE ALASKA REFRIGERATOR CO.

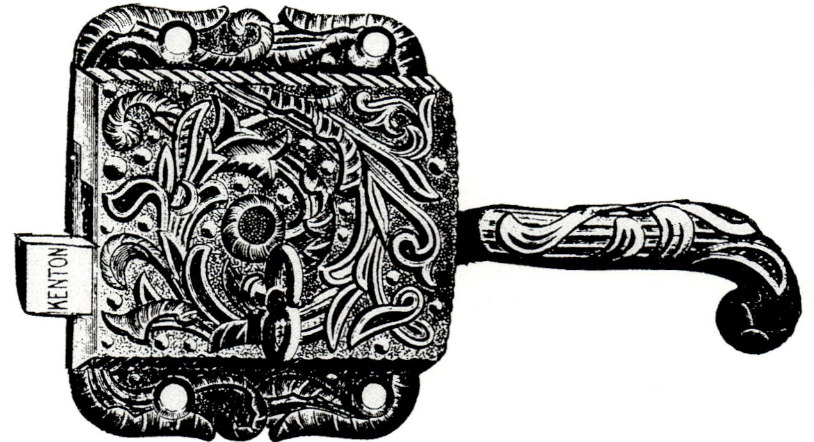

Floyd M. Perkins' patent lock (1892 Kenton catalog).

Keys marked "Kenton," circa 1890-1893. *Courtesy of Hardin County Historical Museums, Inc.*

Real Bronze Mortise Front Door Lock.

3¾ x 3½ Inch. Complete in Sets. Knobs 2¼ x 2¼. Shows Knob No. 100.

With Night Attachments. Cap off Reverse.

One of the many ornate door locks offered in Kenton's 1892 catalog.

No. 1121.

No. 1122.

No. 1124.

Claw and paw feet for furniture (1892 catalog).

Hat and coat hook (1892 catalog).

Drawer pull (1892 catalog).

Perpetual calendar with traces of copper and nickel finish, 1890-1893. *Courtesy of Hardin County Historical Museums, Inc.*

HAT AND COAT HOOKS

To the left we illustrate our hooks as both practical and serviceable; will hold a stiff hat in a proper manner, as well as any other style hat; having hooks for coat and top coat, umbrella or cane.

To the right we illustrate a useful rack for a

hall, clothes closets or wardrobe, by screwing several of our hooks on a board of any desired length and wide enough to accommodate the hook. This line appeals at once to the prospective customer as being a serviceable, beautifully finished hook, and the design right, as a Hat and Coat Hook. Placed on your counters they will sell themselves.

Catalog cut, circa 1920.

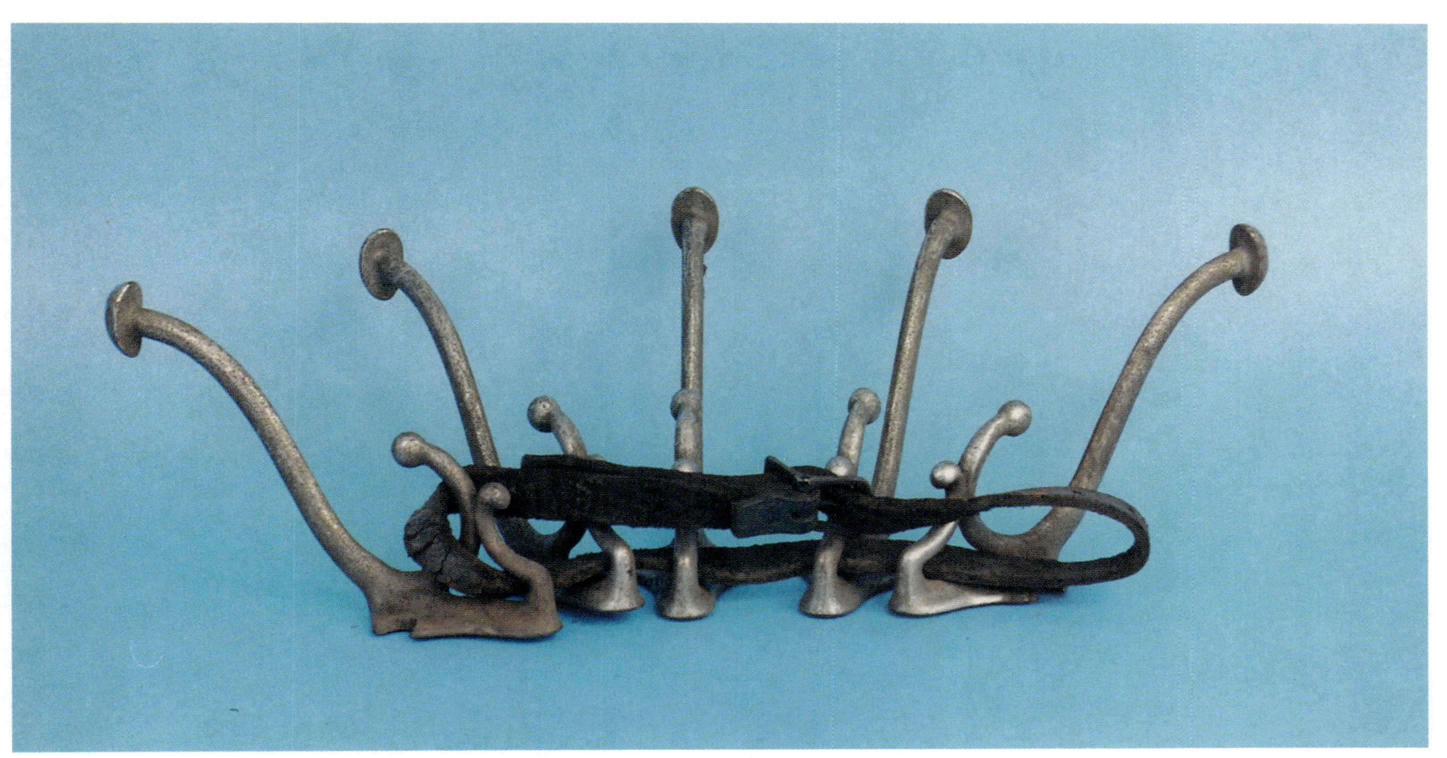

White metal coat hooks on original strap, as found in Kenton stock room.

Chapter 3
World Leaders

As the United States swaggered into a new century, Americans took stock of their burgeoning toy industry. They noted that factories had multiplied threefold in twenty years. They recognized that the total value of trade in toys had grown in tremendous proportions; by the turn of the century, domestic sales were estimated between $15 million and $20 million per annum. America took pride in a business that only a few years before had been called "negligible," but which now offered a serious challenge to European rivals in the world marketplace.

"Iron toys, which we make better than any nation in the world, are sent in large quantities [overseas]," especially to England, noted one journalist of the era. He and other writers observed that the iron toy industry prospered internally as the toy trade learned to depend more upon local production, increased demand, and expanded distribution through drummers (traveling salesmen) and jobbers, mail order catalogs, and department stores.

Ives Toys (Bridgeport, Connecticut), Pratt & Letchworth (formerly Buffalo Malleable Iron Works, in Buffalo, New York), Francis W. Carpenter (Port Chester, New York), William Shimer and Company (Freemansburg, Pennsylvania), Gong Bell (East Hampton, Connecticut), and Harris Toy (Toledo, Ohio) were all seasoned cast iron toy manufacturers. And all, with the exception of Harris Toy, were Eastern operations. The upstart Kenton turned this fact to its advantage, as it had done during the lock manufacturing years. The company's early literature harped on its Midwestern location with "shipping facilities . . . unexcelled, being situated on the Erie, Big Four, and Ohio Central Railroads."

While maintaining its Midwestern base of operations, Kenton was able to broaden its sales network by forging an alliance. At the turn of the century, Kenton joined more than two hundred other toymakers when it contracted with George Borgfeldt & Company, an importer and wholesaler of toys. Borgfeldt, Kenton's sole representative, displayed cast iron toys in a New York sample house or showroom.

The 1900 catalog

Kenton's earliest known toy catalogs, dating from 1900, reveal that the Manufacturing Company joined the ranks of major toymakers in the mid-1890s. Margin notes in one office copy suggest a production date for a line of wheel toys, stoves, and safes as early as 1894.

The 1900 edition describes well over a hundred varieties of cast iron playthings, with road vehicles, including elegant *Hansom Cabs*, *Spider Phaetons*, *Cabriolets*, *Stanhope Gigs*, and *Victoria English Traps*; whimsical *Clown Chariots* and *Dog and Pony Carts*; workman-like *Drays*, *Plantation Ox Carts*, *Log Wagons* and *Ox Wagons*; *New Fire Departments* of horse-drawn fire engines, hook and ladder trucks, and hose carts; "carpet runners" or floor trains featuring an *Empire State Express* measuring an extraordinary 82" in length; "the 'Children's Delight' . . . beautiful and perfect working toy ranges in highly-polished nickel and electro-oxidized finishes" and corresponding *Hollow-Ware* sets of kettles, pots, pans, skillets, and shovels (bearing numbers one through four . . . perhaps some of the earliest, if not the earliest, Kenton toys?); delightful *Steeple Chase* horses on wheels; mules and elephants on skateboard-like pedestals; *Puritan Steamboats* on wheels; two sizes of *Trolleys*; and the venerable "Colonial Design" *Columbia Bank*, in company with "Our New *Elephant Bank*," a *New York National and Chicago Savings Compartment Bank*, and *Home Savings Bank* in electro-oxidized, painted and nickel finishes.

Kenton's rise was nothing less than meteoric. Within a decade of the introduction of the *Columbia Bank* and kindred lines, the concern could boast to the trade, "We are the largest exclusive manufacturer of iron toys in the world." Promotional literature of the new century contained glowing accounts of expanded and strengthened lines—the "most complete on market," "immense storage and manufacturing capacities," and "shipping facilities that are unexcelled." A journalist reported that the company payroll increased from $48,973.67 in 1898 to $72,159.41 in 1902.

The 1901 catalog

"We have spared neither time nor expense to make our line for 1901 larger and better than before . . . our line is particularly strong on fire departments and our stoves are all new and up-to-date designs," began Kenton's follow-up to their 1900 catalog. The "new and up-to-date" stoves included the floridly embossed, nickel-plated and electro-oxidized *Marvel* and *Triumph* deluxe models furnished with tea kettles, kettles, skillets, sauce pans, coal scuttles, and lid lifts. The "mid-range" stoves included *Eve*, *Royal*, and *Rival* with optional black finish and nickel-plated trimmings. In the "economy" line were the *Home, Zoo*, *Ark*, and *Dot* (most appropriately named, for it measured a mere 2.25" wide).

Kenton's fire department nearly doubled in size. Among the notable additions were a *Chief* horse-drawn buggy, two-

and three-horse buggies, two- and three-horse *Fire Patrol* wagons in lengths ranging from 13.5" to 18.5", and a scaled-down one-man *Hook and Ladder*.

Kenton also expanded its lines of banks and wheel toys. The 1901 catalog introduced to the trade a 6"-tall, aluminum-finished *Sky Scraper Bank*; a *House Bank* that became a perennial best-seller; a nickel-plated *Fort Bank*, possibly inspired by the Hardin County Armory in downtown Kenton; a novel *Car Bank* (a cousin to the Kenton *Trolley*); and a *Boom Bank* that eventually became freight for a four-horse *Dray*. Newcomers to the colorful transportation division included a charming *Cupid and Slipper Toy*, a horse-drawn *Trolley*, and a *Red Cross* (2nd Regiment) *Ambulance*.

The 1902 catalog

"[We] have the largest and most complete [toy] line on the market," proclaimed the Kenton Hardware Manufacturing Company in its 1902 catalog. The company pledged that their selection would be even larger and even better than the previous year, that "all lines [would be] greatly strengthened" . . . and the company delivered.

In the spotlight was a *Superior Gas Range*, offered with either nickel or electro-oxidized finishes. It was touted as "a complete and perfect working miniature Gas Range ... something entirely new [that] can be put to practical purposes."

The floor train line was strengthened with a 31.5"-long set comprised of an engine, tender, two flat cars, and crane, and a boxed engine and tender with two dump cars.

One page of the 1902 catalog was devoted to cap pistols. Two of these were animated models, each outfitted with a sliding cannon and an onrushing *Lightning Express* train that doubled as a hammer. More conventional *S&W* "sidearms," plated with copper and nickel, were destined to become Kenton's stock-in-trade. (It is estimated that cap pistols would eventually account for fifty percent of Kenton's production.)

Another page was given to toy cannons in seven different sizes, ranging from a tiny 1 3/8" to a jumbo 9" in length. Customers could choose silvered, gold-bronzed, or japanned finishes.

Making their debuts were the allegorical *Bank of Commerce*, *Bank of Industry*, and *Army and Navy Bank*, each graced with embossed figures and topical motifs. Much more modest, but no less successful, was the economical *Union Bank*.

Certainly new and unique to the Kenton line was a nickel-plated *Rabbit Cart*, a rickshaw-like conveyance. However, the most sensational addition was a line of mechanical toys with a patent device which deactivated the wheels and kept the spring wound when the miniature vehicles were lifted from the floor. Five different designs were offered: a *Hansom*, *Surrey*, *Runabout*, *Steamboat*, *Locomotive*, and *Buckboard*. The last was outfitted with "something new," a spring that wound as it was rolled across the floor and automatically unwound as it reversed on its return to sender.

Together, these offerings made up the "high-grade iron toys" which Kenton had promised its customers, and brought Kenton to a high watermark. In the spring, the company received orders to double its output and to put an expansion plan into motion. Furthermore, a new toy trust–the National Novelty Corporation–offered new opportunities for growth. The company officials rode a wave of optimism and success, and could not have envisioned the cruel fate that awaited them.

"Our *Hollow-Ware* Sets," consisting of miniature polished, nickel-plated, copper, and copper-plated skillets, pots, baking pans, coal buckets and shovels, were listed as numbers one through four in the first known Kenton toy catalog. Does *Hollow-Ware*, and not the *Columbia Bank*, qualify as Kenton's first plaything?

"Special" highly polished cast iron *Hollow-Ware* set. Note that miniature skillet is marked "KENTON BRAND."

The (Admiral) *Dewey Cannon*, circa 1898, commemorated the hero of Manila Bay. Marked "KENTON BRAND," it measures 11.5" x 5.25" x 6". *Courtesy of Mark and Lynda Suozzi.*

Clown Chariot, circa 1900, 10" long. *Courtesy of Bill Bertoia Auctions.*

English Trap, circa 1900, 14.5" long. *Courtesy of Bill Bertoia Auctions.*

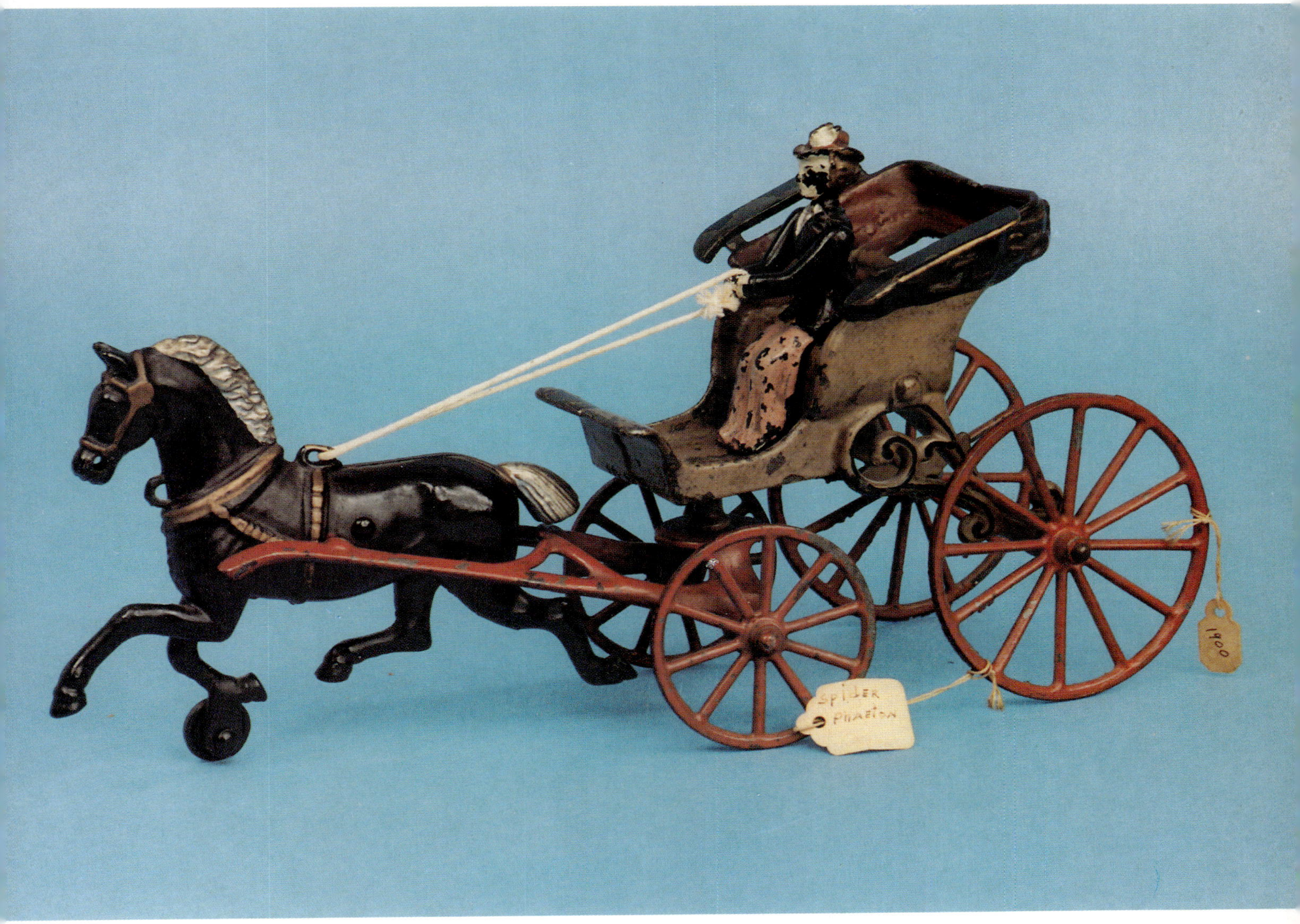

Spider Phaeton, introduced in 1901. This 13.5" long sample room toy has its original tag. *Courtesy of Hardin County Historical Museums, Inc.*

Red Cross Ambulance with Gong, introduced 1902, 16" long.
Courtesy of Bill Bertoia Auctions.

Marvel Range, nickel-plated, measures 19.75" x 16" x 7.5". This item was featured in the 1901 catalog.

Opposite page:
Acme Range, black-enameled and nickel-plated, early 1900s, 13.5" x 12.75" x 6.5".

ACME

New Royal Range, electro-oxidized, 14.25" x 11.75" x 6.5". The mainstay of the Kenton ranges, the *Royal* remained a best-seller for nearly four decades. Early 1900s. *Courtesy of Hardin County Historical Museums, Inc.*

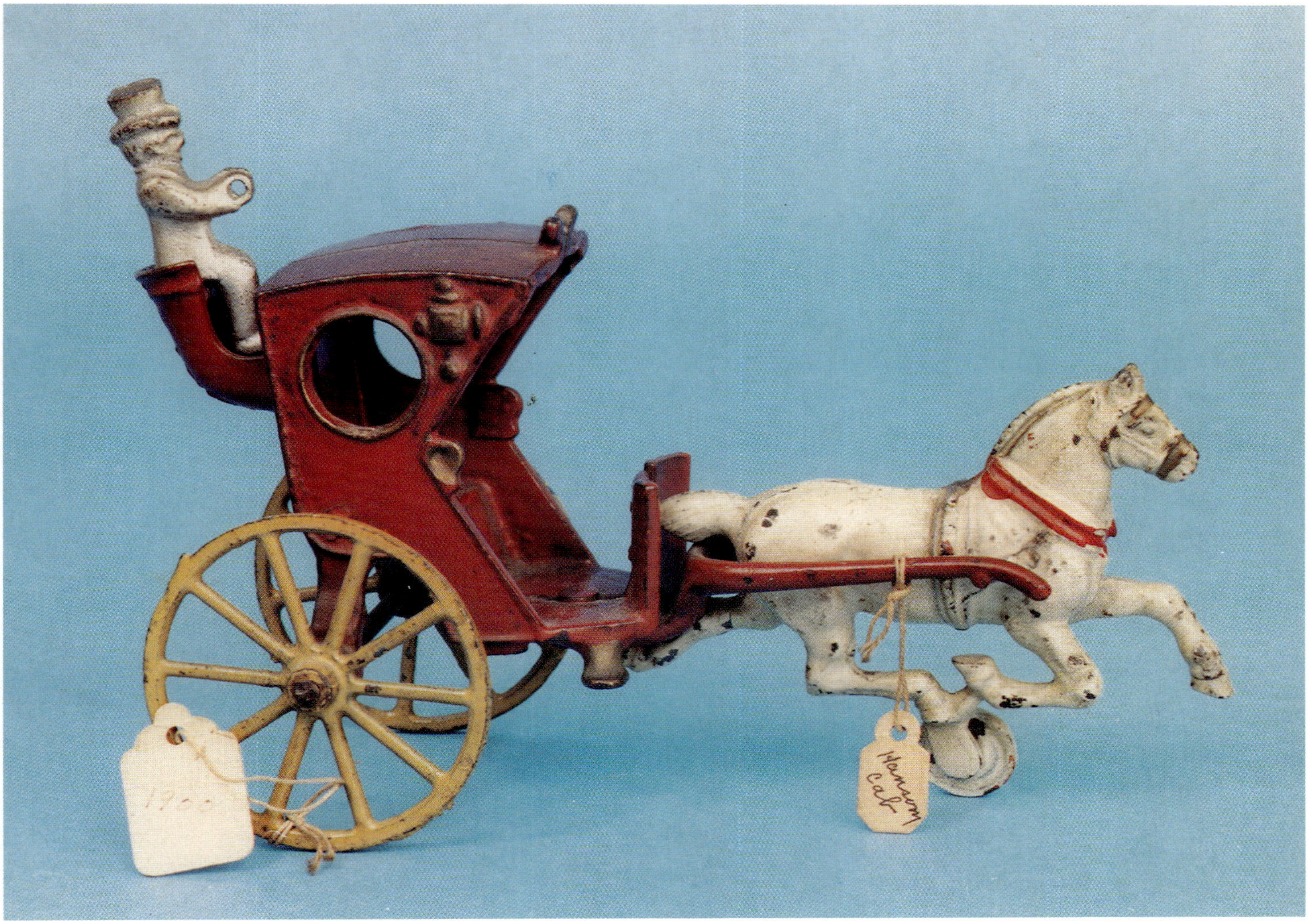

Hansom Cab, circa 1902, is a sample room toy with its original tag, 8" long. *Courtesy of Hardin County Historical Museums, Inc.*

NO. FFF. EMPIRE STATE EXPRESS.

Empire State Express, painted finish, early 1900s. Described as having "extra length," it is an astounding 82" long.

Kenton's *Log Wagon* made its first appearance in the 1901 catalog. This sample room toy with tag marked 1907 is 16" long. *Courtesy of Hardin County Historical Museums, Inc.*

No. GI. OX TEAM AND LOG WAGON.

The *Ox Team and Log Wagon*, 17.5" long, was featured in the 1900 catalog, then disappeared. Described as "A Perfect Model," this toy closely resembles a Hubley product.

Rabbit Cart, nickel-plated, introduced 1903, 4 7/8" long.

"Save Your Pennies" Bank, nickel-plated, circa 1900, 3.5" x 2 5/8".

"Empire Design" *Home Savings Bank*, circa 1900. Painted finish, 5.5" x 4.5" x 3.5". At right is a *Combination Bank*, circa 1900, with painted finish, 4" x 3" x 1.5".

New York National and Chicago Savings Compartment Bank, circa 1900. Electro-oxidized finish, 6.5" x 4".

Top view of a *Compartment Bank*, nickel-plated. This example is inscribed "1896."

House Bank, introduced in 1901, 3.5". This popular, inexpensive model remained in production until World War I.

Union Bank, introduced in 1902. Nickel-plated, 3.75" x 2.5". Another inexpensive and popular model that survived until the mid-1920s. *Courtesy of Hardin County Historical Museums, Inc.*

Boom Bank, introduced in 1902. Nickel-plated, 3.75" x 2.5".

Army and Navy Bank (Double Compartment Savings Bank), introduced in 1903, 6" x 6.5" x 4".

Top of *Army and Navy Bank.*

Bank of Industry, introduced 1902. Nickel-plated, 5.5" x 4.75".

THE KENTON HARDWARE MANUFACTURING COMPANY'S IRON TOYS, KENTON, OHIO. 25

MECHANICAL TOYS WITH NEW PATENT DEVICE.

610. Mechanical Hansom Painted.
615. Same as above, not mechanical.
Length, 6¼ inches.
Height, 7¼ inches.
One in a box; two dozen in a case.
Weight of case, 120 lbs.

620. Mechanical Surrey, Painted.
625. Same as above, not mechanical.
Length, 7 inches.
Height, 7¾ inches.
One in a box; two dozen in a case.
Weight of case, 120 lbs.

640. Mechanical Runabout, Painted.
645. Same as above, not mechanical.
Length, 4 inches.
Height, 3½ inches.
One in a box; three dozen in a case.
Weight of case, 130 lbs.

Our Mechanical Toys all have a patent device which makes it impossible for the wheels to turn and unwind the spring when the toy is lifted from the floor. The above device completely overcomes the greatest objection formerly had to mechanical toys and will be appreciated by the trade.

26 THE KENTON HARDWARE MANUFACTURING COMPANY'S IRON TOYS, KENTON, OHIO.

MECHANICAL STEAMBOAT.

601. Mechanical Steamboat. Painted.
7½ inches long.
One in a box; three dozen in a case.
Weight of case, 120 lbs.

MECHANICAL BUCKBOARD.

650 Mechanical Buckboard. Painted.
9 inches long, 6½ inches high.
One in a box; three dozen in a case.
Weight of case, 120 lbs.
See that spring in front? When rolled across the floor it will return to you. Something new.

MECHANICAL LOCOMOTIVE.

630. Mechanical Locomotive. Painted. 7¼ inches long, 4¼ inches high.
One in a box; four dozen in a case. Weight of case, 120 lbs.

Kenton Hardware Manufacturing Company catalog, 1902.

Locomotive Animated Cap Pistol, sample room toy with original tag marked 1907, 5.5" long. *Courtesy of Hardin County Historical Museums, Inc.*

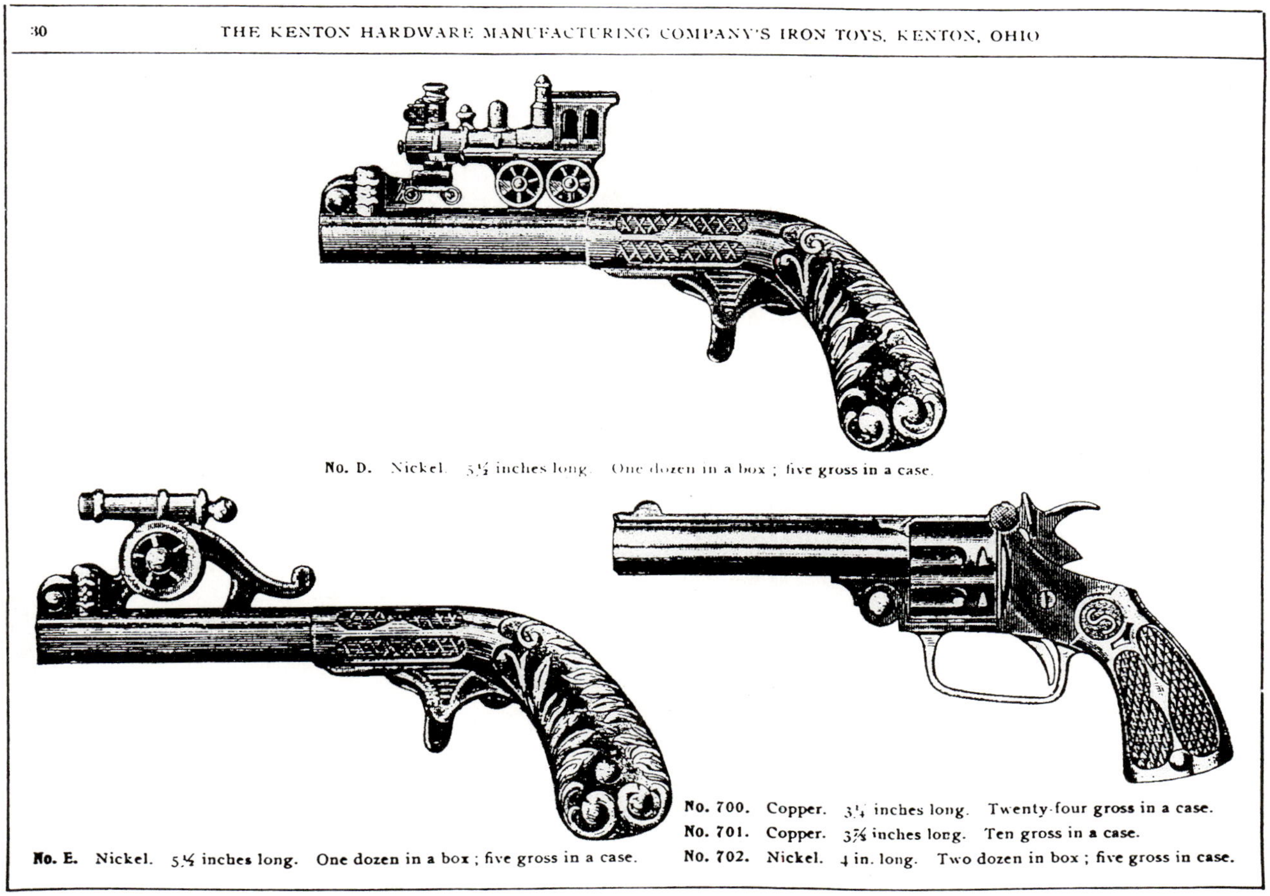

30 THE KENTON HARDWARE MANUFACTURING COMPANY'S IRON TOYS, KENTON, OHIO

No. D. Nickel. 5½ inches long. One dozen in a box; five gross in a case.

No. E. Nickel. 5¼ inches long. One dozen in a box; five gross in a case.

No. 700. Copper. 3¼ inches long. Twenty-four gross in a case.
No. 701. Copper. 3⅞ inches long. Ten gross in a case.
No. 702. Nickel. 4 in. long. Two dozen in box; five gross in case.

Cut from 1903 catalog.

Chapter 4
Fire and Phoenix

The Flames Have No Mercy for Kenton

Another of Our Foremost Industries is Laid in Almost Complete Ruins by the Red Demon

Kenton Visited by a Hundred Thousand Dollar Fire During Monday Night

In a Few Brief Hours the Kenton Hardware Manufactory Almost Completely Wiped Out by a Fierce Fire

And All That Remains is the Foundry of This Industry so Short a While Before a Thriving Institution

Thought to Have Been the Work of a Treacherous Fiend

News of one of the city's greatest conflagrations and most sensational industrial setbacks filled the pages of *The Kenton Daily Democrat* during the week of May 17, 1903.

The blaze started "unusually fast and fiercely" in the centrally-located packing room. In spite of a rapid response and spirited effort from the local fire department, the fire spread quickly and consumed everything but the foundry, and account books and valuable papers rescued from the office. Some $40,000 worth of finished toys were melted into one solid mass.

Rumors flew that a former employee, discharged by the company only a couple of days earlier for drunkenness, was the incendiary, the "treacherous fiend." The assistant state fire marshal arrived within a few days of the conflagration to view what was described as a "blackened mass of smoldering ruins." The findings were inconclusive.

It was ascertained, however, that 170 workers would be put out of work, and that although the company officers had insured their factory for only $52,500, their loss approached $150,000. To make matters worse, Kenton's pending alliance with the National Novelty Corporation had to be put on hold.

A New Toy Trust

The National Novelty Corporation was a toy industry trust that had been recently established in New York, intended to counter what was perceived as the growing menace of foreign imports. The United States had been flooded with tariff-free imported toys, a large percentage of them from the burgeoning industrial giant, Germany. Under the direction of former Kenton superintendent and manager N.H. Colwell, the National Novelty Corporation allied forty-one firms engaged in the manufacture of toys, hardware specialties, and kindred products. The combination followed the leads of the steel, oil, and railway industries in trumpeting the blessings of big business. National Novelty Corporation proponents argued that the trust would introduce "many savings and economies," encourage "thorough cooperation with the trade," and demonstrate that outside competition would prove "neither profitable to the operator nor the buyer."

Before the fire, Kenton had set July 1, 1903, as the date for merger with the trust. Officials of the toy company observed that when it came to comparing the facilities and outputs of the National Novelty Corporation concerns, Kenton "led them all," and that their firm would get even bigger and better through the alliance. Kenton's 1903 catalog, in anticipation of this alliance, bore the National Novelty Corporation cachet.

Kenton's Recovery

The smoke of the smoldering ruin clouded Kenton's future. Company president John H. Pfeiffer, one of the original directors of the lock factory and a grocery magnate, assured distraught employees and investors "the shops will be rebuilt and that such would be done as soon as possible." Within a week, however, he and his fellow officers met with F.H. Harris of the Harris Toy Company and discussed a consolidation and move to Toledo. Additional talks were held with representatives from toy companies in Columbus, Ohio, and Springfield, Illinois. The *Kenton Daily Democrat* reported, "It seems various outside persons and corporations are beginning to show a decidedly worrisome appreciation of the magnitude and worth of the Kenton Hardware Manufacturing Company. The fact that it is the largest manufacturer of toys in the United States is making several towns and factories bid for the factory."

The courters were unsuccessful in their efforts to lure the firm away from Kenton, however. In July of 1903, a local paper reported,

> The Kenton Hardware Manufactory, a phenix [*sic*] of the fire, will be ready for its workers about the middle of August. The force of employees will then be increased to two hundred.
>
> The work of rebuilding was begun as soon as the insurance had been adjusted, and nine masons and two carpenters have been kept busy there most of the time. The buildings are now almost complete. Not only have all the destroyed buildings been replaced with buildings equivalent in capacity and more modern; but a new room 40 by 80 feet has been added above the paint department. The location of the pattern room is changed. The new buildings too, are provided with good ventilation. The walls and doors of the rooms wherein combustibles have been kept have been made fire-proof.
>
> The office is rebuilt, and a new fire and burglar-proof vault, 10 by 6 feet, will be installed. Also three steam elevators will be placed in the factory. Besides this the factory is being equipped with new machinery.
>
> Since the day of the fire, May 18, the moulders have been kept at work. There are now three store rooms stocked with foundry product ready to be finished. Before the fire there were 170 working; since the fire there have been 50; and very soon 200 will be given employment. The Hardware Company is to be lauded for many things.

On January 1, 1904–exactly six months after the date Kenton had originally set for its merger with the "Toy Trust"–the local newspapers reported that stockholders John H. Pfeiffer and W.S. Robinson had just returned from New York, where they had completed the transfer of the Kenton Hardware Manufacturing Company's property to the National Novelty Corporation. As part of the transaction, the Wing Manufacturing Company of Chicago—along with its owner/manager, Peter Wing—would be moved to Kenton and made part of the plant. The move would necessitate an expansion of the foundry and other buildings; the Kenton plant would soon cover an entire city block. A journalist noted that the New Year brought a "new promise of prosperity."

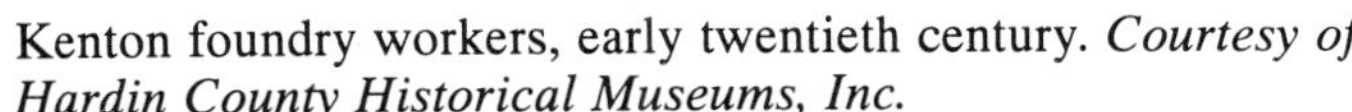

Kenton foundry workers, early twentieth century. *Courtesy of Hardin County Historical Museums, Inc.*

Sulky, early 1900s. Sample room toy with original tag, 8" long.
Courtesy of Hardin County Historical Museums, Inc.

Kenton Fire Department, circa 1900. *Courtesy of Kenton Fire Department.*

Hose Cart, early 1900s, sample room toy with original tag, 17" long. *Courtesy of Hardin County Historical Museums, Inc.*

FIRE PATROL.

P310. Two Horses. Painted Yellow and Red letters. Length. 18 inches. One in a box; two dozen in a case. Weight of case, 165 lbs.

P325. Three Horses. Painted Yellow with Red letters. Length, 18½ in. One in a box; one dozen in a case. Weight of case, 105 lbs.

P510. Fire Patrol same as P310 only smaller horses and painted Blue and Red.

Engraving from an early Kenton catalog.

Chapter 5
Progress and Panic

Theodore Roosevelt, Republican nominee for the presidential contest of 1904, was hailed as a champion of the strenuous life. He was the Rough Rider who would lead America on its quest for world supremacy in commerce and culture. Teddy's name became synonymous with Progressive politics, and his slogan, "Square Deal All Around," appealed to millions of voters. Roosevelt won by a landslide.

Lewis Sharps Bixler, the National Novelty Corporation's newly-appointed general manager of the Kenton Hardware Manufacturing Company, was cut from the same cloth as President Roosevelt. Biographers described him as "a hustler, a moving spirit" . . . "a sympathetic employer with vision and exceptional executive ability" . . . "careful and conservative in business, liberal in his ideas, wide-awake, good-natured, . . . energetic and rated among the moving spirits of Kenton." They added that he was admired for his skill and experience, and that he was the "right man to make a greater success of the enterprise of which he is manager."

The *Greater Hardin County Edition of the News-Republican, the Graphic-News, and the Hardin County Republican* (1910) provided its readers with a thumbnail sketch of Bixler.

> When only fifteen years old he had a position with a toy manufacturing plant (his uncle's, William Shimer & Son) in Freemansburg, Pennsylvania. At eighteen he was promoted to foreman, in which capacity he acted for ten years. He was then made general superintendent, which position he held until 1899, when he resigned to engage in toy and novelty manufacture with C.A. Jones as partner.
>
> Bixler and Jones established their foundry in Freemansburg. The firm continued until the year 1901, when it was incorporated as the Jones & Bixler Mfg. Co., with Mr. Bixler as president and general manager. In July of 1903 the company was merged into the National Novelty Corporation.
>
> Bixler in turn named Michael Wachalec as the new plant superintendent, the successor to Peter Wing, who had returned to Chicago. Wachalec, born in Cleveland, had been with the Kenton concern since its infancy, and was credited with expediting the Lock Manufacturing's transition from builders' hardware to cast iron toy production. Coworkers had regarded Mike as a "man of unusual ability" when it came to making toys, and perhaps just as importantly, an executive who had a gift for "cementing the loyalty of every employee."

The New Catalogs

The first catalog released by Kenton as an official member of the National Novelty Corporation under the Bixler/Wachalec regime reflected some major shake-ups in the toymaking business. Stricken from the pages was the flowery preface of previous editions. In its place was an index that served to impose at least a semblance of order on the ever-changing offerings.

The innovative mechanical (spring-wound) toy line, which had been pruned back the previous year, was deleted altogether. In its stead, Kenton offered novel comic character toys: the spanking *Katzenjammer Toy* and the bowing *Alphonse and Gaston Toy*. The line of wheeled toys was enhanced and expanded significantly with a "New 25¢ Assortment" of painted lion, rhinoceros, boar, bison (buffalo), elephant, and dog carts. Other zoological newcomers included an *Elephant with Canopy*, a facetiously-named *Cairo Express* pulled by an elephant, and a *Cat Cart.* The bank department embraced a new *Flat Iron Building* (inspired by the New York landmark, not far from the National Novelty headquarters) and a (U.S.) *Mail Box.* An important addition to the gun department was the *Rival* "Blank Cartridge only" pistol ("Hole boared [*sic*] through barrel for No. 22 Blank Cartridges. Safety guard to prevent spark from flying in face, and finger guard. Barrel tips down in front to load and eject shells. Big sellers, no line complete without them.").

General manager Bixler recalled that during his first year at Kenton he increased business over fifty percent from the year before, and during the second and third years the increase over the last year of the old management was ninety percent.

In 1905, the company wowed the trade with a number of toys: the *National* (as in Novelty Corporation) *Steel Range*; the *Seeing New York Auto Car* with Happy Hooligan, Mama Katzenjammer, and friends; a *Goat Cart* with Yellow Kid rider; and "Our Ten Cent Leaders," a selection of locomotives, flat cars, and passenger coaches.

In contrast to the lightheartedness of the toys and the warm sentiments expressed in earlier literature, the company offered a no-nonsense policy for the wholesalers:

> On each and every invoice the terms of payment are plainly stamped, and we request ready compliance with the same.... Legal rate of interest charged on all past due accounts. . . . We do (not) guarantee delivery or safe carriage. We do not hold ourselves liable for non-shipments on specified dates.

The year 1906 appears to have been an especially active period for the Kenton branch. Three separate 1906 catalogs for cannons and pistols, toys, and sad irons were released. In addition to the cast iron playthings, Kenton continued to promote novelties and hardware castings in brass and bronze and electroplating in nickel, copper, and oxidized copper.

The cannons and pistols were advertised as "the most complete [line]–the workmanship and finish unexcelled." The company literature proclaimed that "Our Pistols work the best and produce (the) loudest report," and the safety-conscious were assure that "All our Pistols have extension(s) on hammer to prevent spark(s) from flying in face." Three different versions of a *Smith & Weston* [*sic*] pattern pistol (or "pattern *pistlo*," as the catalog description had it) with imitation rubber grip were illustrated; two (one priced at five cents, the other at ten cents) were described as "easily mistaken for a 6 shot revolver." Three pages of the Kenton catalog were devoted to Stevens' Patented Toy Cap Pistols that were made in both the Kenton and Stevens factories, each a part of the National Novelty Corporation. ("The NATIONAL guarantees the workmanship and finish of this line. NUFF SED.") A corresponding document indicates that pistol orders for west of Buffalo and Pittsburgh were to be shipped from the Kenton branch. Patriotically painted cannons (3.25" to 9.5" in length, priced to sell at ten cents to fifty cents), with red wheels and blue frames, were just right for "young America."

Toy sad irons and stands, ranging in price from $7.78 to $37.38 per gross, were featured in polished, gold-bronzed, and nickel-plated finishes. Patented detachable wood-handled designs, including the celebrated *Mrs. Potts* model, remained popular Kenton products into the 1930s. Freight allowance schedules and tariff tables frm circa 1910 to circa 1930 indicated that sad irons received significantly lower shipping rates than the general toy line, and thus offered the opportunity for greater profits.

As a sideline, the Kenton cast, polished, and plated miniature meat pounders ($8 per gross), hatchets ($9.50 and $10.50 per gross), and tack hammers ($4.25 and $4.50 per gross). A schedule of selling prices for 1907 Kenton products listed a variety of garden tools, bombs, hooples, sand toys, jacks, picks, boys' reins, and roller chimes–all products not found in the standard catalogs.

The Panic

At roughly the same time, in domino fashion, several major banks collapsed, stocks plummeted, and thousands of businesses failed. Millions of Americans were caught by over-extended credit. The result—"The Panic of 1907."

By spring of 1907 it became apparent that Kenton was carrying a lot of dead weight, and managers from both the local branch and the National Novelty Corporation headquarters approved the wholesale disposal of old stock. On the "hit list" were venerable varieties of the *Columbia Bank*, *Dog Cart*, and *Hook and Ladder* wagons. The Wing product line was decimated as nearly three dozen banks and ranges were dropped from production.

Bixler's office copy of the 1907 catalog, "with corrections and notations," reflects the tumult. Most of its pages are defaced with scratch-outs, the rubber-stamped "CANCELLED," and the handwritten "off" and "out." This wholesale makeover is further substantiated by a ledger that reveals that National Novelty Corporation president N.H. Colwell authorized the discard of approximately one hundred Kenton and Wing wheel toys (including the *Sightseeing Auto Car*, *Cupid and Slipper Toy*, and *Fire Patrol* models), banks (among them three sizes of the *Columbia*), ranges, and sad irons.

To offset the discard, Kenton emphasized its automotive designs ("We are leaders in this line") and showcased "up-to-date" *Expresses, Electric Hansoms, Tourist Cars,* and *Red Devil* automobiles.

In addition, Kenton entered the crowded (and, as it turned out, depressed) mechanical bank market with two designs, the *Teddy Bear Bank* ("When money is put in the Bank the Bear opens its mouth") and *Mrs. Katzenjammer Bank* ("Mrs. Katzenjammer rolls her eyes when money is put in the Bank"). The former followed on the heels of New Yorker Morris Michtom's wildly popular plush, stuffed bruin. The latter attempted to capitalize on the popularity of the Kenton Katzenjammer wheel toy. Both newcomers proved to be spectacular failures. Kenton carried them for two years, then opted to end the experiment.

A Toy Trust Turnover

In response to the instability, the parent National Novelty Corporation tried to assure its customers that they would be taken care of in "the most satisfactory manner." The Shimer & Son branch at Freemansburg proceeded to close, and in mid-November 1907, L.S. Bixler told anxious Kentonites, "Except for a few days or possibly a week for our annual inventory, we will remain open all the time."

The week after this pronouncement, however, brought news of the demise of the National Novelty Corporation—along with the cancellation of orders and a two-month shut-down of the foundry.

Under the direction of a second toy trust, the Hardware and Woodenware Company, Kenton (one of sixteen branches) resumed operations, but business was shaky. Once again, the product line underwent radical surgery; dozens of models of ranges, *Columbia Banks*, Wing *Columbus Banks*, wheel toys, can openers, tack hammers, and hat and coat hooks were written off as discards.

Another Shut-Down

Approximately one hundred and fifty employees went to pick up their paychecks on February 8, 1908, and found this notice posted: "We are in the hands of a receiver." N.H. Colwell, president of both trusts, was now receiver; his counsel told *The Cincinnati Enquirer*, "The company's embarrassment was due to the contraction of its bank credits during the recent financial crisis."

Kenton shut down operations once again, and inventory was taken. Colwell left his New York office and returned to Kenton, his former home, where he told bankers that $20,000 had to be raised or the branch could not resume work and Bixler would begin dismantling the plant. Kentonites were told they stood much to forfeit: a facility appraised at $150,000, with $30,000 of annual sales in Fourth of July goods alone.

In an appeal reminiscent of the fund-raising campaign that brought the original lock concern to Kenton, the citizens were called upon to rally in support of the toymaker. The Kenton Commercial Club, proclaiming that the rescue of the operation was in the best interest of the public, headed up a fund-raiser. However, it was apparent that the people of Kenton had become prejudiced against the company since it had become part of a trust, controlled by outsiders, and the necessary funds were not forthcoming. When asked about the future of the Kenton Hardware Manufactory Company, Bixler had nothing to say except that he had been shipping patterns out to other Hardware and Woodenware factories. (The Kenton toy line would be continued by other trust branches.) Before long, Bixler himself was gone, having been appointed the trust's superintendent of ironworking branches. He took plant superintendent Michael Wachalec with him.

President L.S. Bixler in the Kenton office, early 1900s. *Courtesy of Hardin County Historical Museums, Inc.*

Plant superintendent Michael "Mike" Wachalec in the Kenton office. *Courtesy of Hardin County Historical Museums, Inc.*

Superintendent Mike Wachalec (with bowler, at center) and Kenton employees outside the plant, early 1900s. *Courtesy of Hardin County Historical Museums, Inc.*

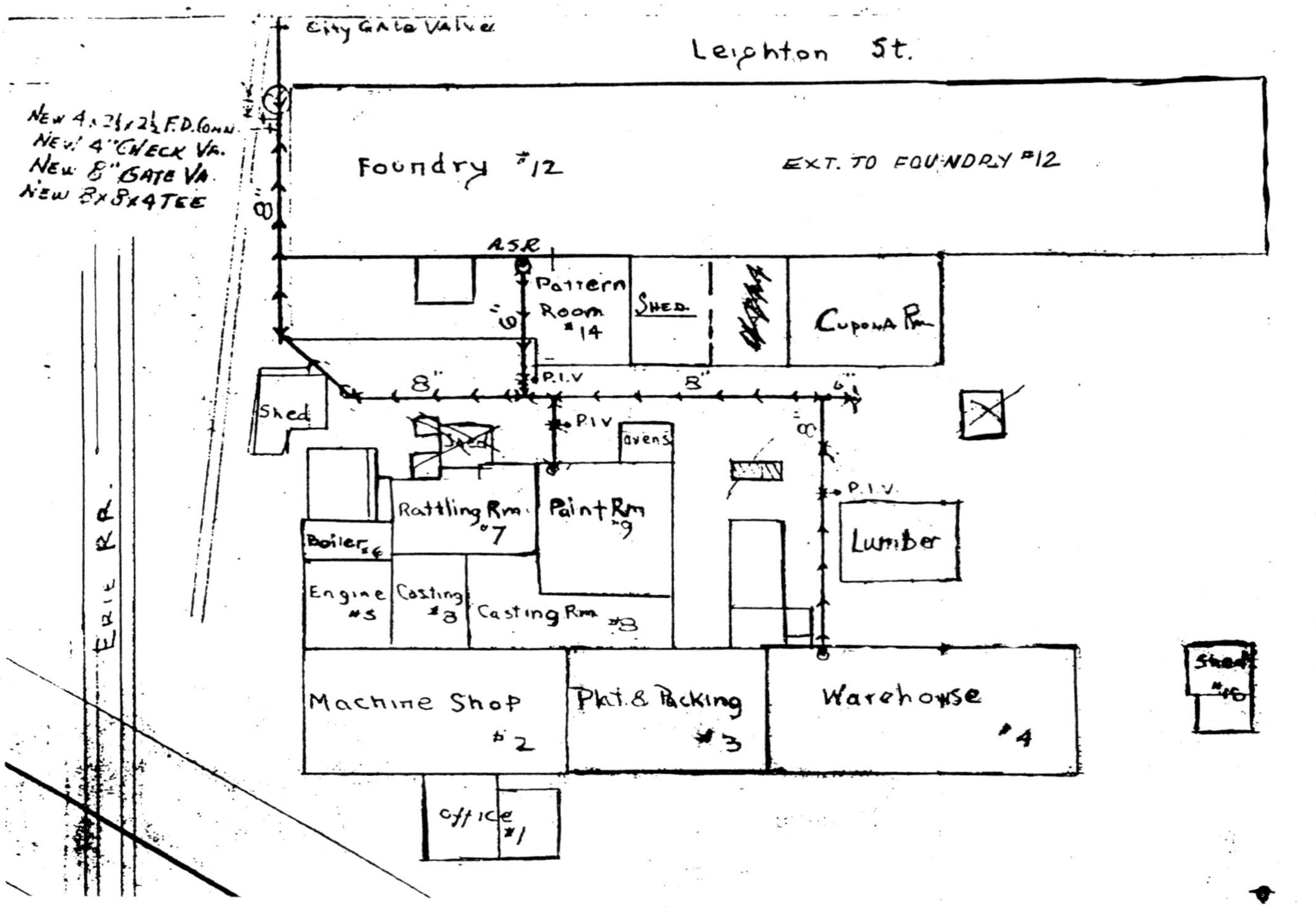

Recently discovered floor plan of the Kenton factory, early twentieth century. *Courtesy of Dan Allen.*

Plant operations, early twentieth century: molding with fine sand and nonferrous (brass and soft white metal) patterns connected by gates. *Courtesy of Hardin County Historical Museums, Inc.*

"Rattling Room." Parts were tumbled and made sand-free in drums on left, then quality-checked and sorted at tables on right. *Courtesy of Hardin County Historical Museums, Inc.*

"Bombers" (issued honorary "Pilots' Licenses") grinding flash from parts. *Courtesy of Hardin County Historical Museums, Inc.*

Polishing, tumbling, and buffing. *Courtesy of Hardin County Historical Museums, Inc.*

Riveting. *Courtesy of Hardin County Historical Museums, Inc.*

Assembling toy stoves. Plant superintendent Michael Wachalec is in bowler. *Courtesy of Hardin County Historical Museums, Inc.*

Seeing New York Auto Car, introduced 1905. "The $1 size," 10.5" long. *Courtesy of Skinner, Inc.*

Katzenjammer Toy, introduced 1905. The "50 cent size" without rear rider, 12" long.

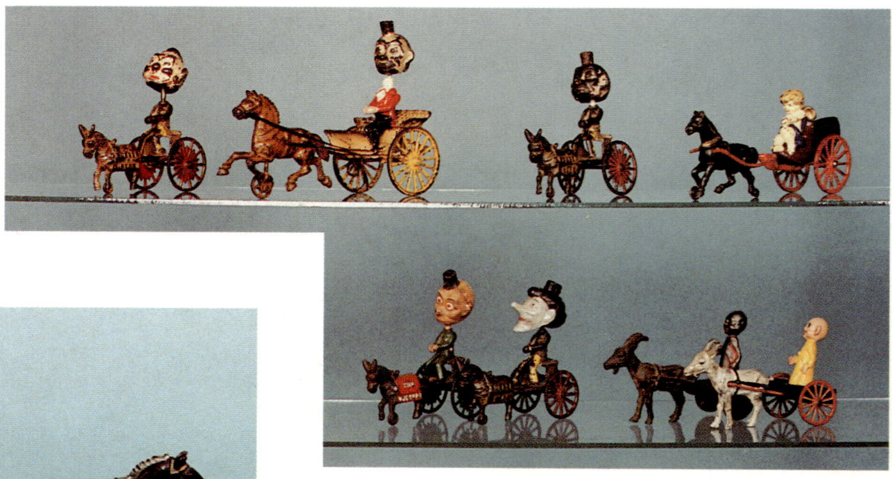

Top, left to right: *"The Nodders" Half Dollar Nodding Head/ Nodder* (Gloomy Gus), 6.5" long; *Large Nodder* (Happy Hooligan), circa 1910, 11.5" long; *"Rubber Neck,"* circa 1910, 6.5" long; *Cupid and Slipper Toy*, early 1900s, 9" long. Gloomy Gus and Rubber Neck may be Hubley models; note the characteristically ornate wheels. Bottom, left to right: *"The Nodders"* (Happy Hooligan), circa 1910, 6.5" long; *"Rubber Neck"* (cartoonist Fred Opper's urbane Alphonse), circa 1910, 6.5" long; *Goat Carts* (Yellow Kid, America's first comic-strip character, created by Richard Outcault for William Randolph Hearst), circa 1910, 7" long. It has been speculated that the driver was painted black and his nightshirt red after Kenton lost the Yellow Kid contract.

Happy's Patrol, introduced in 1905, 17.5" long, with wheel-driven mechanical action. Cartoonist Fred Opper's Gloomy Gus drives the Patrol, while his brother, Happy Hooligan, is rapped on the head by an officer of the law. *Courtesy of Bill Bertoia Auctions*

Katzenjammer Toy, 12" long. As wheel turns, Mama spanks child. *Courtesy of Skinner, Inc.*

Orange version of *Seeing New York Auto Car* (restored), 10.5" long. *Courtesy of Bill Bertoia Auctions.*

Egyptian Assorted Toys, introduced 1904, 8" long. This "25 cent line" enjoyed a long production run. The *Elephant Cart* is a later variation.

Egyptian Assorted Toys. Early catalog engravings show the drivers with reins.

Elephant with Canopy. A toy from the National Novelty era, 5.25" long.

Cat Cart. Sample room toy with original tag, 1904, 6" long.

Left: *Rabbit Cart*, 4 7/8" long. Right: *Cat Cart*, 6" long, circa 1910 to circa 1915. Later versions of these wheel toys were offered in "colored assortments." *Courtesy of Bill Bertoia Auctions.*

Cairo Express, introduced 1904, 9.75" long. This facetiously named cart was long-lived. The elephant was reincarnated as a bank and as novelties (see Chapter 9). *Courtesy of Bill Bertoia Auctions.*

Mrs. Katzenjammer Bank (mechanical) as pictured in 1907 catalog. A variation with nude children is rumored to exist.

Teddy Bear Bank (mechanical) as pictured in the 1907 catalog.

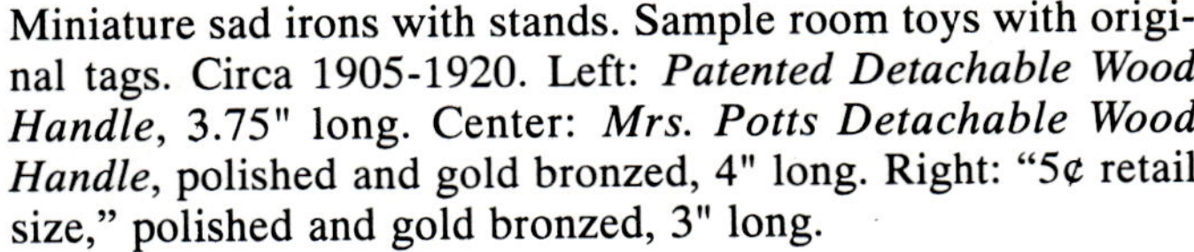

Miniature sad irons with stands. Sample room toys with original tags. Circa 1905-1920. Left: *Patented Detachable Wood Handle*, 3.75" long. Center: *Mrs. Potts Detachable Wood Handle*, polished and gold bronzed, 4" long. Right: "5¢ retail size," polished and gold bronzed, 3" long.

Polar Bear Bank (mechanical) from the sample room. Believed to be a unique prototype. *Courtesy of Hardin County Historical Museums, Inc.*

Left: *Teddy Bear Bank* (mechanical), 1907, "painted the color of a Cinnamon Bear" (later models electro-oxidized), 7" x 3.5". Right: *Teddy Bear Bank* (mechanical), a sample room toy assembled from nonferrous patterns. *Courtesy of Hardin County Historical Museums, Inc.*

Sample room toy, 3.5" long, "the 3¢ size." A similar toy gun is shown in the Kenton Toy Pistols and Cannons Catalog of 1906. *Courtesy of Hardin County Historical Museums, Inc.*

Cannons, early 1900s. Left: Pattern toy with nickel-plated brass barrel, assembled by Kenton employee Harold "Buddy" Spencer, 9" long. Right: Model painted with remnants of original "patriotic" red and blue paint, 4.5" long. *Courtesy of Hardin County Historical Museums, Inc.*

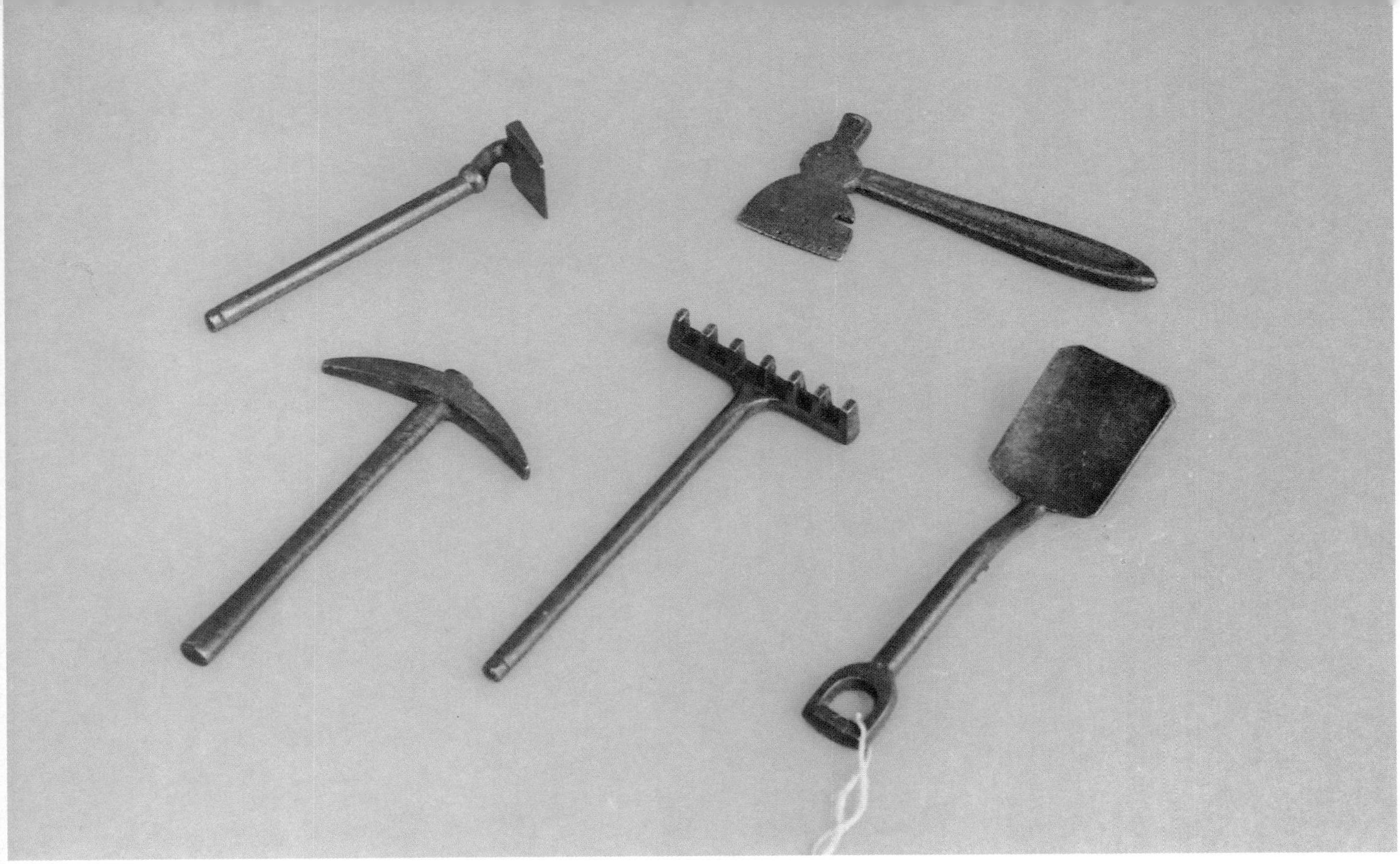

Small penny toys attributed to Kenton. Found in the walls of a Hardin County, Ohio, home with a *Teddy Bear Bank* and other Kenton cast iron products. The Arcade Manufacturing Company of Freeport, Illinois, made similar articles with nickel-plated finishes.

Chapter 6
"A Sounder and More Permanent Basis"

"Mr. L.S. Bixler, for many years President of the Jones & Bixler Company and later Manager of the Kenton Hardware Mfg. Co., has become actively identified with our company," announced the Hubley Manufacturing Company in a 1908 issue of *Playthings* magazine. Bixler had resigned his post with the trust. With Wachalec as his associate, Bixler looked to guide Hubley, a victim of a major fire, on an ambitious rebuilding and expansion program.

Within a year, though, Bixler was back with the Hardware and Woodenware Company and back in Kenton. The Kenton Hardware Manufacturing Company was reborn.

"Hardware Plant to be re-opened . . . Joyful news for Kenton . . . Mr. Bixler to Return." "HARDWARE PLANT TO BE RE-OPENED . . . JOYFUL NEWS FOR KENTON . . . MR. BIXLER TO RESUME MANAGEMENT," trumpeted front page headlines of the November 15, 1909 edition of the *Kenton Daily News*. Bixler told reporters that the receivership would be lifted by the first of the year. In addition, he assured *Daily News* readers,

> . . . the company is being reorganized on a sounder and more permanent basis than ever before and . . . local people may have every confidence in the re-establishment of the Kenton Hardware Manufacturing Co. As some indication of this you can see that Mr. Wachalec and myself would not have resigned good positions at Lancaster if we had not felt confidence in the permanency of the plant here after it is open.

Two weeks later Wachalec had returned to Kenton and directed the clean-up of the plant. He expedited the return of patterns that had been sent to other Hardware and Woodenware branches. Plans called for the factory to be up and running by January 1. He expected he would have to start 'slack' (slowly), because of the loss in trade.

Business papers of the early part of 1910 reflect a prompt tautening of sales policy:

> To *strictly* jobbers only; an extra 5% discount (on minimum and regular case quantities) to be allowed, also To Retailers and recognized syndicates (5 & 10¢ syndicates not included) on $1000 worth and over our general Holiday lines of Iron Toys (Wilkins, Jones & Bixler, Grey Iron, Stevens, & Kenton Branches) an extra 5% discount to be allowed.
>
> No direct or indirect concessions to be made as an inducement to secure any order of these Iron Toys.
>
> No orders to be accepted at regular case quantity prices or minimum case quantity prices in less than the prescribed case quantities.
>
> No deviation from regular packing to recognized syndicates.

A Dizzying Array of Toys

Whereas sales policies constricted, the toy line blossomed. The first general catalog published after the reorganization, an eighty-two page tome, offered a dazzling array of ranges, wheel toys, fire departments, pistols, and banks. The firm took special pride in the *Happy Foxy Nodding Toy* ("Gloomy Gus driving and Foxy Grandpa and Happy Hooligan nodding their heads at one another. The decorations are the best ever put on a Toy. Length, 16 inches."), *Uncle Sam Nodding Toy* ("Uncle Sam Nods his head as Foxy Grandpa drives the horse. A beauty."), *Half Dollar Nodders* ("Beautifully Decorated"), and *Nodding Head* ("Rubber Neck") *Assorted Toys*. By 1910, Kenton listed no fewer than eighty horse-drawn fire wagons; a year later, the toymaker added the *Auto Fire Patrol*, *Auto Hose Wagon*, *Auto Fire Engine*, and *Auto Hook and Ladder*. Bixler's office copy of a 1910 catalog suggests an experiment to outfit former combination banks with keyed locks.

Kenton Supremacy

Manager Bixler, with a renewed optimism and zeal for his work, was on a mission to reclaim for Kenton its supremacy in the cast iron toy trade. He took a direct hand in the development of the products, and in 1911 was awarded a patent for the ornamental design of a bank (the *Safety Vault*, which appeared in the catalog of the same year), acquired the exclusive rights to manufacture and market an innovative *PAP-ER-KRAK Toy*, described as "The Sane Noise Maker ... No Fire, No Smoke, No Danger, All Noise ... AMMUNATION [*sic*]–OLD NEWSPAPER ... Best working and Most Reliable air torpedo on the market." (In 1913 Bixler himself would design and patent *Paper-Exploder*, an improvement on the *PAP-ER-KRAK*.) The old Kenton standby, the *Columbia Bank*, was revived, cast in four sizes (4.5" to 9" high) and plated in nickel and electro-oxidized finishes. The hardware line was likewise resurrected; corporate records document the production of no fewer than twenty-one varieties of hat and coat hooks, as well as a selection of shoe lasts.

Nearly four dozen jobbers, from New York City to Spokane, Washington, from Montreal, Canada, to Meriden, Mississippi, appeared on the Kenton roll. By January of 1912, the price schedule for toys ran thirteen pages long and listed nearly five hundred varieties. Thanks in large part to Bixler and Wachalec, the fortunes of the Kenton branch had been reversed dramatically within the span of two years.

Another Breakup?

Just when the prospects began to look bright for the concern, rumors began to circulate about the pending doom of the Hardware and Woodenware Company. The last Kenton catalog prepared under the combine's name, a 1912 supplement, was attached to the previous year's full-line catalog, and together they appear to have served as the "standard" 1912 catalog. It must not have been an encouraging sign for the employees or the customers.

The Hardware and Woodenware Manufacturing Company advised its jobbers, "We make no allowance for breakage [of our goods in transit]." Within months, however, it did have to make allowance for the breakup of the toy trust.

President L.S. Bixler's patent papers, found in an office safe years after the factory closed.

Safety Vault Bank, front and back, patented 1911. Nickel-plated, 6.75" x 5.5". *Courtesy of Hardin County Historical Museums, Inc.*

Safety Vault Bank, electro-oxidized, 6.75" x 5.5". Lacks front grate. Sample room toy with incorrect tag.

Patent design for *Iron Clad Bank* designed by Bixler.

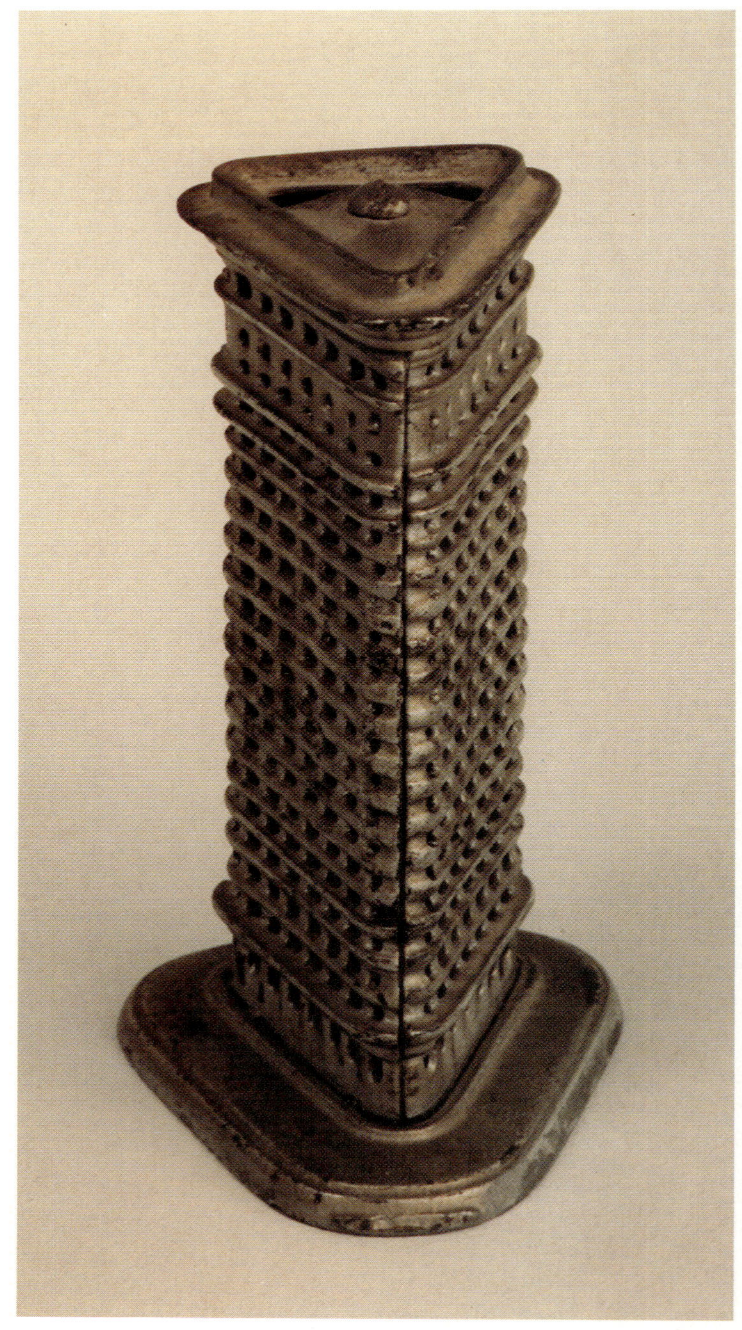

Flat Iron Building Bank, aluminum finish, 6" high. Based on the famous New York landmark, this popular toy, introduced in 1904, was produced for nearly a quarter of a century.

High Rise Bank, aluminum and gold finish, 5.5" x 2 7/8" x 2 7/8". Sample room toy with original tag marked 1909. Often attributed to Kenton, but not listed in contemporary catalogs. Could it have been made by another Hardware and Woodenware branch? *Courtesy of Hardin County Historical Museums, Inc.*

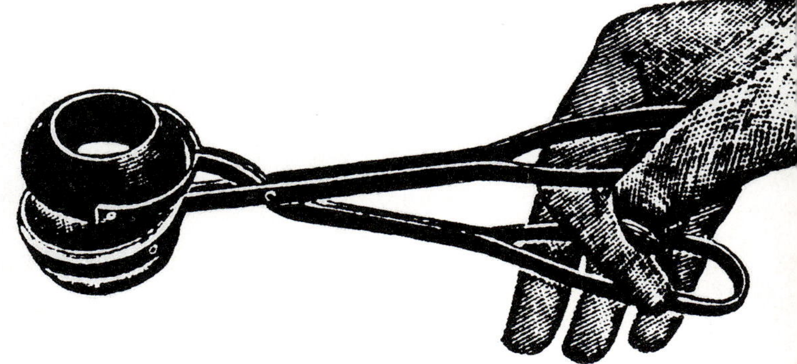

Open to receive paper

Closed ready to explode

The *PAP-ER-KRAK* in action, as illustrated in a Kenton catalog.

The *Globe Safe*, 5.25" high. Left: a unique specimen with highly polished nickel-plating, custom made for employee Harold "Buddy" Spencer. Right: electro-oxidized sample model. Although tagged and dated in 1907, the *Globe* did not make its appearance in catalogs until after the 1910 reopening. *Courtesy of Hardin County Historical Museums, Inc.*

Passenger Train, nickel-plated, circa 1910, 27" long.

Road Cart, 7.5" long, sample room toy with tag marked 1908.
Courtesy of Hardin County Historical Museums, Inc.

Sulky, sample room toy with original tag marked 1909, 5" long. *Courtesy of Hardin County Historical Museums, Inc.*

Coal Cart, 7" long. Inexpensive mule-drawn vehicles were part of the Kenton stable from the trust years through World War I.

Early automobile line, circa 1910. Left: *Locomobile*, 5.5" long; center: *Red Devil*, 4.25" long; right: *Automobile*, 7" long.

Chapter 7
Incorporation

In April of 1912, Hardin County, Ohio (like the rest of the world) was overwhelmed by the news of the *Titanic* disaster. Many Kentonites might have missed a news item that related the demise of another giant, the Hardware and Woodenware Company.

A.H. Tisdale, representing Bixler and fourteen other branch managers, purchased the assets of the toy combine for $440,000 at a receiver's sale at the courthouse in Worcester, Massachusetts. Bixler told the press,

> This means the liquidation of the company and the independent plants of the company throughout the country. It means that twelve of the managers of the fifteen plants, most of whom were former owners of the business, bid for the plants and will become their owners. I am making arrangements for the taking of the Kenton plant and through the assistance of Eastern friends have raised enough of the funds to finance two-thirds of the purchase price of the Kenton factory. The balance I hope to raise in Kenton by interesting local capital in the enterprise.

In August of 1912, the Kenton Hardware Company was incorporated under the laws of the State of Ohio. According to the articles of incorporation, the business was "formed for the purpose of manufacturing articles of iron, steel, wood, rubber, and other materials, and transact all business incident thereof." L.S. Bixler, Mike Wachalec and three other prominent Kenton businessmen were listed as its incorporators.

Re-establishing Kenton

The transition from the Kenton Hardware Manufacturing Company to the Kenton Hardware Company was hardly smooth or seamless. Business accounts for 1912 reveal a gaping hole between the dissolution of the trust and the incorporation. Kenton had been part of a combine since 1904, and there was a task of disassociation—of distancing the concern from the ill-fated Hardware and Woodenware Manufacturing Company and severing the ties with most of its former affiliates.

One branch, however, the Jones & Bixler Manufacturing Company of Freemansburg, Pennsylvania, was to re-affiliate itself with Kenton; by the end of 1912, Lewis Bixler had engineered the purchase of his old firm. He brought its patterns, equipment, and stock (as well as his brother, W.H. Bixler) to Kenton. Jones & Bixler continued producing cast iron toys under its own name while using the Kenton facilities until 1914. Then the company dissolved, and its miniature scales, fire departments, passenger and coal trains, wagons and carriages, automobiles, and banks were dropped or mainstreamed into the Kenton line.

By the year's end it was clear that Bixler and Wachalec had made great strides in re-establishing Kenton as the leader in its field, certainly meeting, if not exceeding, the expectations of their investors. Two weeks before Christmas 1912, a feature writer for the *Kenton News-Republican* toured the Kenton Hardware Manufacturing Company and came back with a glowing report. He headlined his feature, "KENTON KNOWN IRON TOY CENTER–Greatest Iron Toy Manufactory Plant in the World Located Here."

"Santa Claus must have placed large and liberal orders with the Kenton Hardware Company," the article began,

> for many carloads of iron toys had been sent to all parts of the world. . . . some little Japanese lad will be happy Christmas morning with seeing his toy train making good time or a small Philippine girl will try her hand on cooking on a genuine American stove of stovelet size. . . .
>
> And [the toys] went to several other countries also. There is not a state in the Union in which the Kenton make of iron toys are not found. The Kenton Hardware Manufacturing [*sic*] Co. is now the largest maker of iron toys in the world, making the greatest number of any concern in the country.

The reporter went on to walk the reader through a factory tour.

> The starting point is the pattern room. An idea is born of some one. It must be a splendid idea. It must be something that will catch on with the trade. It must be something that will sell for after all selling is what tells in the manufacturing business. It may be a pistol, a vestibuled train, a fire engine, a range, a bank, but it must have the merit of utility, novelty, finish, originality or individuality. Suppose it be a pistol of a new shape. The man with the idea imparts to the pattern-makers. The pattern-makers get busy and make in wood a pattern of each part of the pistol that when assembled would make a whole pistol. These patterns are made of wood. There are many thousands of them in this company's plant. They are arranged and classified in the pattern room. The next natural place is the foundry room, where the castings are made. The men who work here are moulders. They have a sand, fine grained firm that stays where it is put. An impression is made from the pattern in this sand and the moulten metal is poured in through a hole in the tip. It cools and hardens, there is your casting. If it is a small part several may be cast in one mould. Following our pistol as illustrated we may say that we now have one half or side of a pistol. Then it goes to the polishing room. Many hundreds of them are thrown into revolving drums. With them are thrown hundreds of little fine pointed devices that go tumbling about with the castings and in the end by friction the same and such particles are removed. Then the pistol is assembled in another room, that is the two parts, the trigger, etc. are fastened. Then if a silver pistol is wanted they go to the plating room and are dipped in an electric bath where anode and cathode create a current that causes deposits of silver to seek their affinity to another metal and they find their way to the iron of the casting. Then more polishing, in more drums and finally to the finish room. But there are many and various other ways of finishing and so there is oxydizing, enameling, annealing, polishing and painting all very interesting in their several ways. One of the finest finishes is the copper and gun metal. With cast iron as it is in wood the better the casting the more readily it lends itself to a perfect finish. The Kenton Hardware Manufacturing [*sic*] Co. prides itself on the finish of its toys. They are as highly polished, as carefully finished as are the real articles made by the leading manufacturers. Their toy safes for example are as carefully finished as would be the big ones of the most representative safe builders.
>
> When the toy is fully finished it goes to the packing room where nimble fingered girls put them in cartons ready for shipment to their destination.

The *News-Republican* reporter wrapped up the feature by taking comfort in knowing that the firm was once again a "home institution . . . controlled by men of this city . . . interested in the enterprise [and with] faith in the plant."

He concluded,

> The volume of business has been extremely good. The plant has been made and more and more are to follow. A new water tank is being erected and a deep well pump is to be installed.
>
> L.S. Bixler is the general manager of the company . . . He knows the business thoroughly and he is putting plenty of energy and ability in the working of making the business grow and he is succeeding.

The years before the First World War must have been exhausting, even for the energetic general manager.

Woodrow Wilson and New Struggles

The new U.S. president, Woodrow Wilson, promised to abolish any and all protective duties, "everything that bears even the semblance of privilege or any kind of artificial advantage." The resultant Underwood Tariff Act opened the door for foreign competitors, and by 1914, approximately half of the domestic market belonged to Germany and other alien rivals. Without the luxury of protection and a supportive combine (Wilson also opposed trusts) Kenton girded for what looked to be a grueling campaign to keep its market.

Kenton's business ledger documents the assimilation of the Jones & Bixler toy line, the company's expansion of its jobber network and the incursion into new markets (e.g. playsuit retailers), the sweeping changes in discount policies and freight schedules, and the radical pruning of old stock. In 1914, dozens of traditional toys were struck from production, among them all sizes and finishes of the *Columbia Bank*, the *Electric Hansom*, six varieties of the *Ice Wagon*, nearly half the line of automobiles, the *Clown Chariot*, the *Cairo Express*, and the *Happy Foxy Nodding Toy*. A note in the ledger suggests that the toymaker looked to unload a whole line of hat and coat hooks in 1916.

Between 1913 and 1914, Kenton had dropped the selling agent George Borgfelt & Company in favor of the Riemann, Seabrey Company of New York. Organized in 1908 by George P. Riemann Jr. and Edward William Seabrey, the upstart firm had used fresh marketing tactics to attract the leading toymakers to their fold. "We specialize in service," avowed the direct factory salesmen of the Riemann, Seabrey Company. They informed the toy buyer,

> Our service begins with a careful selection of Toy Lines that are unmistakably par excellence.
>
> We display them in our Permanent Toy Fair for your convenience. You are spared time, trouble, traveling, and the accompanying expense by seeing so many under our roof. Our service does not end with the taking of your order. We follow up your shipment. We supply you with electros [electrotypes], catalogs and price information concerning our lines.
>
> In short, we are the salesmen of each manufacturer we represent. You buy at factory prices here.

While dozens of cast iron toys were dropped from the Kenton line, dozens more were introduced to the market. Executives told customers, "We are constantly adding to our catalogs . . . Our line is the most complete–the workmanship and finish unexcelled . . . Our slogans:—'GOOD GOODS,' 'SUPERIOR QUALITY' and 'PROMPTNESS IN FILLING ORDERS.'" They called "especial attention" to new lines of floor trains, ranges, pistols, *PAP-ER-KRAK Toys, Sane Exploders, Boy Scout Outfits,* and belts and holsters. A 1913 Spring Goods catalog spotlighted a new model *Smith & Weston Revolver Blank Cartridge Pistol.* The 1915 supplement featured six varieties (20.5" to 38") of an *Erie-Vestibule-Limited-Train*; new "Kenton" line *Steel Gas Stoves* ("up-to-date . . . complete in every detail–burns gas . . . Beautifully nickel plated . . . Made in 4 sizes"); three sets and nine individual pieces of *Hollow-Ware* (tea kettles, boilers, skillets, and sauce pans) in three finishes (copper-plated, polished and nickel-plated, or highly polished and nickel-plated); a "beautifully painted and decorated *Auto Water Tower* . . . 24" long, 24" high tower is raised"; and two additions to the *Boy Scout Outfit* line, nickel-plated *Smith & Weston* cap pistols packaged with imitation black patent leather belts and holsters.

Once again, L.S. Bixler took an active role in toy design. On November 7, 1915, he was awarded two patents for the ornamental designs of single- and double-burner gas stoves. By the end of the winter, the United States Patent Office granted him additional certificates for closed and open "magazine hammerless patterns," improvements which were incorporated in the innovative *Boy Scout Cap Pistol* and *Automatic Target Model.*

Cannons and cap pistols, which had been once marketed as "Fourth of July goods," became all-season playthings. Florid Victorian designs with leaf and scroll embellishments and figural mechanisms gave way to realistic guns, with plainer exteriors but more intricate interiors. Kenton devoted more and more of its time, attention, and resources to cap pistols. Eventually, approximately six months of each year were set aside for their production. The years 1916 and 1917 saw exceptional activity: Kenton purchased the William Shimer, Son & Company pistol department, further expanding its line. Kenton would prove itself a formidable rival for the Stevens, Hubley, and Kilgore cap pistol powerhouses.

Kenton and World War I

Indeed, by 1914 Kenton once again billed itself as the "largest factory in the United States and Canada" devoted exclusively to the cast iron toy business. Continued growth and prosperity resulted in a stock split. Kenton's domestic counterparts became increasingly conciliatory as the decade progressed: the pressure of foreign competition brought them closer together, and World War I and the resultant slump in worldwide toy sales and the restrictions on production materials united them. President Woodrow Wilson created a regulatory War Industries Board to conserve the fuel and minerals so vital to the toy industry.

In anticipation of the pinch, forty-one toymakers (Kenton among them) formed a group called the Toy Manufacturers of the United States of America. In June 1916, L.S. Bixler attended the Manufacturers' organizational meeting in New York and was named to its board of managers. Bixler, with a "who's who" of the toy industry which included A.C. Gilbert, A.F. Schoenhut, and H.C. Ives, issued a statement to the press:

> [The aims of the Toy Manufacturers of the United States of America are] to raise the whole toy industry to a higher ethical standard, to cooperate and help the toy merchants in every way and to carry on an organized movement to push the sale of American-made toys not only in this country but throughout the world as well. . . .

The organizers stressed that unlike the previous National Novelty and Hardware and Woodenware combines, the new association was not formed for the purpose of regulating or controlling prices.

As their overseas markets were by and large closed, the manufacturers looked to preserve their industry from the spartan policies of War Industries Board chairman Bernard Baruch. While it looked discouraging, toy production was on a day-to-day basis; Kenton's 1917 Schedules of Selling Prices were qualified with this special notice: "Deliveries cannot be guaranteed, but will be subject to factory's ability to procure labor, material, and supplies." Fletcher Dodge, secretary of the Toy Manufacturers, reported, "The cost of raw materials has been going up and they have been in fact difficult to get at any price." The metal toymakers are "backward on deliveries," he added. In addition, the Hardware Company advised, "Prices subject to change without notice."

However, in the end, the allies of the toy industry were successful in their campaign against the Allies of war. Domestic toymaking would go on in spite of the global conflict. The manufacturers were able to convince the powers that be that Americans needed cast iron playthings. And besides, they argued, Christmas wouldn't be the same without Kenton.

Gas Stove, black paint with nickel-plating, 6" x 6" x 4.5". The "New 'Kenton' Line . . . A Practical Toy Which Burns Gas." *Courtesy of Hardin County Historical Museums, Inc.*

Steel Gas Stove, black paint with nickel plating. Ornamental design patented by president L.S. Bixler on November 9, 1915.

Mail Box Bank, aluminum finish, 3.75" x 2 5/8". Part of the Jones & Bixler line, 1912-1913.

Liberty Bank, 9.75" x 3.5". Sample room toy with original tag, circa 1915. Initially a Wing product, later a Jones & Bixler offering, and finally a Kentontoy.

After 5 days return to
THE KENTON HARDWARE CO.
KENTON, Hardin County, OHIO

MANUFACTURERS OF
The "KENTON" Line
TOYS

SAFES
BANKS
TRAINS
RANGES
SAD IRONS
WHEEL TOYS
CAP PISTOLS
AIR TORPEDOES
AUTOMOBILE TOYS
PAP-ER-KRAK TOYS
FIRE DEPARTMENTS
BLANK CARTRIDGE PISTOLS

We do not hold ourselves liable for non-shipment on specified dates, nor do we guarantee delivery or safe carriage of goods. All shipments F. O. B. Factory.
ALL DELIVERIES CONTINGENT UPON STRIKES, ACCIDENTS OR OTHER CAUSES BEYOND OUR CONTROL

THE KENTON HARDWARE COMPANY
KENTON, OHIO

IMPORTANT
MAKE ALL CHECKS PAYABLE TO
THE KENTON HARDWARE COMPANY
AND MAIL TO KENTON, OHIO

INV. NO.
YOUR ORDER NO.
OUR ORDER NO.
TERMS
DATE
SOLD TO
ADDRESS
SALESMAN ORDER NO. SHIPPED VIA NO. PKGS

QUANTITY	DESCRIPTION	PRICE	EXTENSION	TOTAL

Rare Kenton stationery postmarked 1913. Receipt for plating auto parts signed by president L.S. Bixler. *Courtesy of Hardin County Historical Museums, Inc.*

Scales, sample room toy with tags marked 1913. Left, 3.5" x 1.75" x 1.5" (without scoop). Center, 6" x 3.5". Right, 5.25" (without scoop) x 3" x 3". *Courtesy of Hardin County Historical Museums, Inc.*

Toy Fence, sample room product, painted green with gold decorations, sections total eight running feet. Introduced circa 1912. *Courtesy of Hardin County Historical Museums, Inc.*

Cat Face Ashtrays, made by Kenton; gilt and nickel-plated. *Courtesy of Hardin County Historical Museums, Inc.*

Art Nouveau style ashtrays embossed "COMPLIMENTS OF THE KENTON HARDWARE CO.–IRON TOYS." Polished cast iron and nickel-plated, 7" x 6.5". *Courtesy of Hardin County Historical Museums, Inc.*

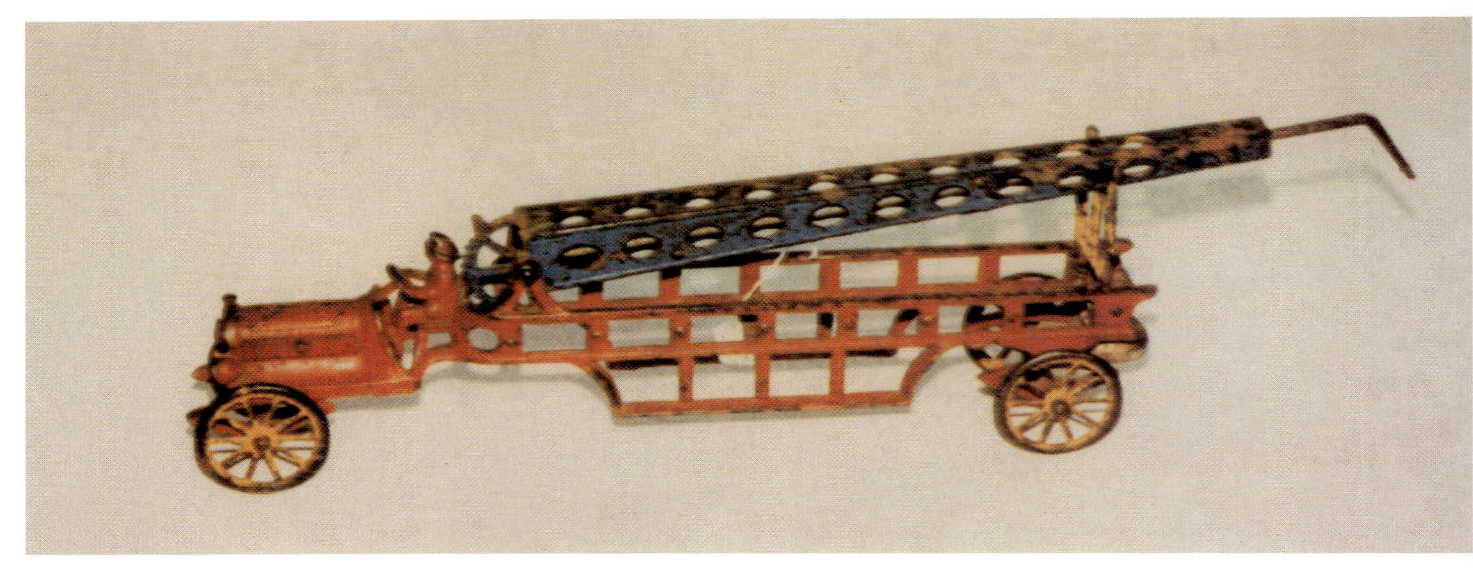

Water Tower Truck, circa 1916, 24". *Courtesy of TOYS etc.*

"Flag raising at Kenton Hardware Mfg. [*sic*] Co.," 1917. Tradition has it that the wooden pole was one of the tallest in the United States.

'Bixy, Big Boy'

L.S. Bixler was recognized by his fellow members of the Toy Manufacturers of the U.S.A. as the dean of the cast iron toy business. In his community he was frequently recognized for his civic-mindedness and benevolent spirit. The Kenton employees likened him to Santa Claus . . . and at one company function, he actually dressed the part! In the company's prosperous years, he shared the wealth, providing his workers with low-rent housing, life insurance policies, medical benefits, a "100 Percent Club" for sharing and airing concerns as well as enjoying company fellowship, year-end bonus checks and gifts, inscribed loving cups recognizing years of distinguished service, and Christmas banquets.

At one such banquet, employees hailed him as "the greatest president of the greatest industrial plant of the greatest town in the world." In a radio-style production, the "Unknown Sextette" sang to the tune of "Charlie, My Boy":

Bixy, Big Boy, Oh! Bixy, Big Boy,
You scare us but share us
your pleasures and joys.
You've got the kind of stuff
that brings us success.
In all the ups and downs of bus-i-ness.
And when you're away,
We'd all like to play,
But we're all quite certain
this wouldn't pay.
They tell me you are quite the fisherman, too,
We wonder if all your stories are true,
And when you catch one, I've heard you say two.
Oh! Bixy, Big Boy!

Two other principals in the Kenton success story were not spared. Willard, the president's son and assistant manager (soon to become vice president) was serenaded.

Willard, My boy, Oh! Willard, My Boy . . .
The toys you design
Are certainly fine
And each year you add a
lot to the line . . .

Plant superintendent Mike Wachalec was informed,

Michael, Me Boy, Oh! Michael, Me Boy . . .
You've got that kinda, sorta, bit of way,
That makes us work, and yet it seems like play.
Out in the shop, we labor a lot,
But Michael, it's worth it,
As boss, you're a top.

Cartoon of downtown Kenton, Ohio, circa 1930. President L.S. Bixler approaches the Hardin County Courthouse. *Courtesy of Hardin County Historical Museums, Inc.*

Loving cups presented to superintendent Mike Wachalec and foreman Claud Conner by president Bixler during Kenton Hardware Company's annual Christmas banquet, 1929. *Courtesy of Hardin County Historical Museums, Inc.*

Chapter 8
Boom & Gloom

The War to End All Wars did not end war, but it did check Europe's domination of the toy industry. At the outbreak of war, Germany, England, France and other foreign powers accounted for more than seventy-five percent of the toys sold in American stores. By 1920, the tables had turned and imports fell to an estimated ten percent. The American toy industry grew fivefold to keep pace with domestic demand. Only Japan, which had experienced comparable growth, remained to challenge the United States for supremacy. By 1927 American toy manufacturers supplied ninety-five percent of domestic demand.

"We lead the world in toys," proclaimed William Fleming French of the magazine *Illustrated World*. Writing in 1919, French boasted that American toymakers were not only leaders in sales, but in ingenuity and invention as well. He added that the back of foreign competition had been broken, and a new decade would promise great opportunities for Americans in overseas markets.

There was another encouraging sign for the trade: more and more prominent educators across the nation came forward to advocate the design and manufacture of "real use and value" for children. The educators noted that the war had revolutionized and Americanized the toy market, creating more and more demand for toys that were modern, realistic, and functional.

Kenton was in the catbird seat. After the growing pains, the ups and downs of the first two decades of the twentieth century, the company had matured into a major toymaker with the resources and vision for continued ascendancy in the trade. Promoters affirmed that years of experience had honed the company into an efficient and enthusiastic guild. Their staples– the "models right up-to-the-minute in design and finish" . . . "exact miniatures of those things in daily use" . . . "masterpieces of cast iron construction"–dovetailed with the conception of an ideal toy espoused by educators, what L.S. Bixler referred to as "toys that teach."

Bixler and his associates pledged that theirs were the "NEWEST AND BEST IN IRON TOYS . . . attractive numbers, well made, finely finished in bright colors, handsomely decorated . . . 'KENTONTOYS' with all the name implies."

The sunny outlook was briefly darkened by an eclipse. American industrialists had underestimated the resilience of the Germans and the tenacity of the Japanese, and by the early 1920s these rivals looked to regain dominance in America's domestic toy market. The federal government, spurred on by the Toy Manufacturers of the U.S.A., intervened with stiff tariffs, blocking the foreign threat and protecting Kenton and its fellow American toy manufacturers. From 1921 to 1927 the Bureau of Foreign and Domestic Commerce quadrupled its appropriations to American business. Toy imports plummeted from a high of $8,362,437 in 1923 to a low of $4,057,754 in 1926, while exports nearly doubled in the same time period.

According to catalogs of the post-World War I era, Kenton's up-to-the-minute designs were matched by up-to-date equipment. The factory was expanded to encompass over two acres of floor space, and the staff swelled to over a hundred administrators, secretaries, pattern-makers, foundrymen, painters, platers, riveters, assemblers, and sheet metal workers. Company literature continued to hail the Kenton Hardware Company as "the largest factory in the United States or Canada devoted to the exclusive manufacture of iron toys." The toys represented the "highest quality made," unexcelled "for cleverness in design, for artistic decoration, and for beauty of finish," according to company literature. Distribution reached an all-time high, and Kenton stocks split again in 1921.

These were indeed the Roaring Twenties for the Kenton Hardware Company. Business records reveal an incredible flurry of activity between 1919 and 1929, and the New York sales office of Riemann, Seabrey was supplemented by agencies in Boston and San Francisco. Young & Glenn, Inc., was established as an export office, and Kenton participated in the New York, Chicago, and San Francisco toy shows.

Kenton did a thriving business, with no fewer than seventeen New York syndicates with networks of retail stores across the United States and Canada. A seemingly endless stream of crates poured out of the plant via rail and road, and the toymaker filled a Great Lakes freighter. Among the more prominent establishments which sold Kentontoys were Lord & Taylor of New York, the Higbee Company of Cleveland, the F&R Lazarus Company of Columbus, Ohio, L.S. Ayres & Company of Indianapolis, The Vogue of San Antonio, the Bon Marche of Seattle, and James A. Ogilvy's, Ltd. of Montreal.

By 1923 Kenton offered a staggering seven hundred basic varieties of toys, and countless variations with custom finishes. Among the varieties was a new line of cap pistols, "revamped and changed . . . to obtain the maximum of spring efficiency to insure a power spring action, and . . . [with] added new protective features." Small pistols with small names (*Rex, War, Bat, Hio, Imp, Art, Pup, Dix, Dox, Jax,* etc.) and small prices proliferated. Long-barreled guns with longer names (*Western, Kit Carson, Buffalo Bill, Wild West*) profited from luxurious finishes of nickel, buff, and copal varnish.

The highlights of the 1924 transportation toy line were the new *Auto Taxi, Ford Sport Cars,* and the *Auto Circus Calliopes* and *Cages*. The *Cages,* designed and patented by L.S. Bixler, featured a red paint scheme for a lion, white for a bear, and yellow for a hippo.

At the Riemann, Seabrey Company's "Permanent Toy Fair" at 11 Union Square West in New York City, Kenton unveiled "46 new numbers of iron toys of the latest design and most modern equipment" on February 2, 1925. Throughout the year the firm ballyhooed a new line of gas ranges in royal blue and nickel, an automotive fire department ("actual reproductions of the present equipment in the largest cities"), a *5th Avenue Bus* and *Bus Coach* with "New Type Balloon Tire Wheels," four new models of *Moving Van*s, three *Sport Racer*s, two *Tank Truck*s, and one *Radio Bank* with a two-dial combination lock (a three-dial model joined the lineup two years later).

"A series of beautiful automobile toys–*Sedans* and *Coupes*, accurate in every detail, both in design and finish" was the highlight of 1926. The automotive chassis was cast separately from the body, allowing the "true reproduction of wide-crowned fenders–and beautiful body proportion." The color schemes mimicked those seen at the Paris Automobile Salon. "No other toys at the same price show as much service and satisfactory value as iron toys," sales literature touted, adding, "THE BEST ARE 'KENTONTOYS' . . . There is no substitute for Quality."

The Kenton Hardware Company made news headlines during the 1927 Christmas season when it contributed toys to children in flooded areas of New England. Among those gifts were a new array of *Fire Engine*s, *Extension Hook & Ladder*s, *Police Patrol*s, *Tank Truck*s, *Speed Truck*s, *Dump Truck*s, and a new line of stoves ("embodying the present mode–simplicity with richness–strictly quality finish . . . real quality at popular prices").

In 1928, President L.S. Bixler told a reporter that "many changes have taken place [in the past twenty-five years of toymaking], as has been the experience of all lines of business–a veritable evolution in the goods manufactured, the methods of merchandising, selling, and every other angle conceivable in bringing the toy industry up to its present status." He indicated that the toy lines of the early 1900s were "primitive" in comparison with those of 1928, adding that toys had to change frequently with constant improvement in finish and stability in construction to "hold their place to-day with the trade."

Implicit in Bixler's synopsis was a growing concern for the welfare of his industry and the anticipation of difficult challenges that lay ahead. By 1930, sales of cast iron toys dipped to an estimated two to five percent of the retail market, while bicycles and other wheel goods, dolls, and mechanical toy sales grew, taking some of cast iron's market share. A domestic economic decline coupled with tariff reductions and stiffer competition from foreign manufacturers exacerbated the situation. Kenton's domestic rivals, notably Arcade and Hubley, responded by making toys that were "snappy" and "stylish." The slogan for Hubley toys was "They're Different"; the Lancaster, Pennsylvania company produced such models as automobiles with detachable bodies and electric lights, *Lifesavers Truck*s, *Harley-Davidson Highway Patrol*s, *Beach Patrol* boy surfer, and a *Daddy Long Legs* grasshopper toy that chirped as it was pulled. With these toys, the company probably established a design edge on Kenton. Arcade, with its equally clever designs and innovative, eye-catching packaging with "color, color and more color!" pulled ahead of the Kenton Hardware Company. Kilgore, pioneers of plastic technology, made significant inroads into Kenton's cap pistol market.

A review of price schedules from 1927 onward suggests that Kenton attempted to become more competitive by reducing minimum quantities required for discounts and cutting the prices of toys themselves. Bixler and his associates began to streamline and modernize their toy line, phasing out many traditional ("primitive") models of floor trains, combination banks, ranges, sad irons, cap pistols, fire departments, and other horse-drawn wheel toys. At the beginning of the 1920s, Kenton advertised five hundred models; by the end of the decade, that figure was halved. Kenton literature dropped references to general hardware, plating services, and, finally, iron novelties.

There were dramatic changes in store for Kenton. In 1929, a watershed year, the firm introduced "Hot Numbers" with "snappy colors" that were promoted with the newly-adopted slogan, "The Real Thing in Everything But Size." The "Hot Numbers" (the *Graf Zeppelin, Pickwick Nite Coach, Passenger Train, Electric Range, Air Mail Plane, Trailer Assortment, and Tractor*) were incorporated into a new, playful logo.The busy, cluttered graphics of previous catalogs and advertisements were replaced with slick art deco layouts. The *Pickwick Nite Coach* was "the most unique and graceful design on wheels." The "newest thing" in toy electric ranges were models equipped with "a real switch that clicks when turned on and off and raises the 'red hot' heating element into 'cooking position' ... just what the modern child as well as the modern mother wants." The *Graf Zeppelin*, which was guaranteed to "go over big with young and old alike," and five sizes of *Air Mail Planes* cashed in on the aerial craze of the era. Likewise, mechanical "action" toys were all the rage and the *Jaeger Concrete Mixer*, "A Good Mixer on Any Toy Counter," appealed to "young engineers." Also new to the Kenton line were "up-to-the-minute" boxed assortments that offered the retailer the "fine opportunity to sell four toys instead of one."

The Crash

"CHEER UP!" the editor of *Playthings* told the toy trade shortly after the stock market crash of October 1929. Willard R. Bixler, son of president L.S. Bixler and a vice president himself, followed this advice. He assured his co-workers and clients, "The outlook for the Toy Industry . . . should be better than ever. If developing and marketing new toys will help the situation any, we feel we will be doing our share. We are looking forward to a good year." Bixler took special pride in additions that had a "New Zip and Zest as modern as 1930."

Among these were *Horizontal and Vertical Engines* which were closely modeled after large engines used in mills, cotton gins, and power plants. Retailers were advised that a 1/8" belt could be cut from an old inner tube and used to connect the engines to an electric motor; "Try this stunt in your window for a quick 'sell out' . . . also stress this feature in selling these toys." *The Buckeye Ditcher* (the "sensation of 1930"), was inspired by a product made in nearby Findlay, Ohio. It was outfitted with two cranks that engaged chain-sprocket drives—"intricate, yet very easy to operate." A *Flying Boat* with a large pusher-type engine and revolving propeller was another modern-style creation.

Nonetheless, Willard Bixler was evidently disappointed by sales. He observed that improvement was "bound to come," and cautioned that it had to be "built up slowly and surely." During the spring of 1931, three new trucks (*Contractors*, *State Highway*, and *Lumber*) were said to have created a sensation at the Chicago Toy Fair. However, the toy expected to be the real blockbuster for the year was the *Morgan Crane*. Contemporary admirers described it as perhaps the finest and most interesting Kentontoy ever produced. In addition, it was probably their most intricate and elaborate product. Sold "knocked down" in an attractive carton, the *Morgan Crane* measured 16" by 12.5" by 12" when secured to a wheel base. A wheel on a control car and two levers governed three-way movement of the crane. By lifting a fourth lever, the operator could release a load (the deluxe model came with a locomotive and tender) suspended by a hook. The *Morgan Crane* proved to be one of Kenton's biggest fizzles, indicated by the very few examples that appear on the current antique toy market.

The Kenton-Kingsbury Connection

The Bixlers came to realize that new products alone would not and could not recapture Kenton's lost glory. They saw progress in partnership, and formed an alliance with the Kingsbury Manufacturing Company of Keen, New Hampshire, in 1931. L.A. Carll, formerly of the Riemann, Seabrey Company, was appointed sales manager. Carll pronounced that the "new setup" (partnership) meant greater service to the toy trade. The Kenton-Kingsbury promotional literature reasoned,

> The business of the Kingsbury Manufacturing Company and the Kenton Hardware Company do not conflict: they complement one another. The Kenton Toys are constructed of cast iron. Kingsbury Motor Driven Toys are built of steel, while the [Kingsbury] *Silver Arrow* flying planes are constructed of wood and aluminum. The elder Mr. Kingsbury and the elder Mr. Bixler are among the original toy men in the United States and both are still active in their resptive [*sic*] businesses. The new sales policy brings these two successful companies into active cooperation and will enable dealers to obtain a complete line of quality toys and popular toys, sold at attractive prices through one sales channel.

President C.L. Kingsbury elaborated that the affiliation "will enable us to maintain a closer and more direct cooperation with dealers and will provide, under one selling plan, a complete line of cast iron toys that provide everything the dealer wants in service, profit and good-will." Added L.S. Bixler, "Kingsbury and Kenton . . . have long enjoyed friendly relations and it is most natural, that, in order to improve their sales-service to dealers, they should combine their selling efforts."

Kenton and Kingsbury shared a two-page spread in *Playthings* and adopted coordinating graphics.

In truth, the partnership was a desperate act in desperate times. The industrial-production index (an economic barometer comparable to the gross national product) for 1929 reached 119; by 1932 it had slipped to a mere sixty-four. Conversely, unemployment skyrocketed. Millions of Americans made do without three square meals, let alone toys.

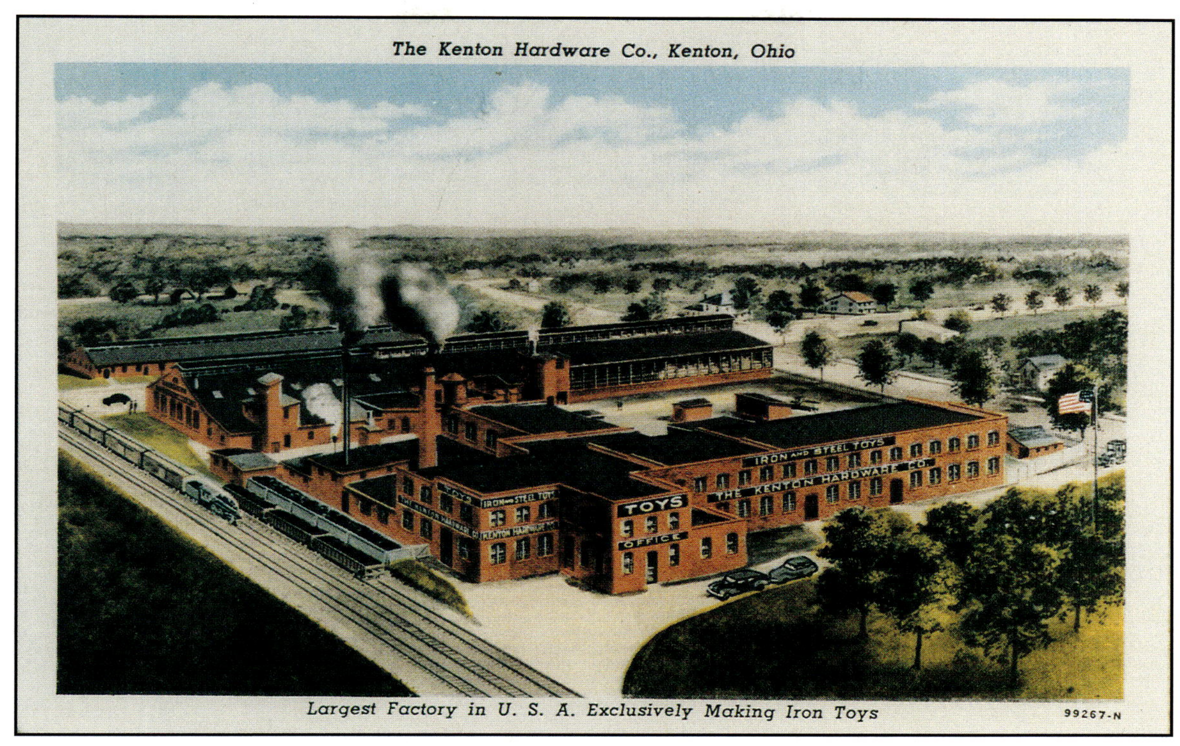

Postcard view of the plant, circa 1925. *Courtesy of Hardin County Historical Museums, Inc.*

Promotional brochure, circa 1920.

NO. 516 AUTO

Painted and Decorated. 6½ inches long. 3 in a box. 6 dozen in a case. Weight per case, 150 lbs.

40 Numbers of Automobile Toys

All kinds and designs of Autos, old and new. We are adding new numbers to this line constantly, furnished either with disc or spoke wheels as you may choose, Touring Cars, Sedans, Auto Fire Toys, Coal Wagons, Cabs, Drays, Patrol Wagons, every kind of Automobile Trucks.

NO. 807 AUTO DRAY

Painted and Decorated. 15 inches long. 1 in a box. 2 dozen in a case. Weight per case, 225 lbs.

Why you should buy—

"KENTONTOYS"

RIEMANN SEABREY COMPANY

11-15 Union Square, West, New York, N. Y.

Direct Representatives for U. S. and Canada

The Famous

"KENTONTOYS"

Are manufactured by the KENTON HARDWARE CO., Kenton, Ohio, U. S. A., the largest factory in the United States or Canada devoted to the exclusive manufacture of iron toys.

There are 500 numbers in the line, representing the highest quality made.

For cleverness in design, for artistic decoration, and for beauty of finish, they are unexcelled.

Their attractiveness makes them sell on sight. A few numbers are reproduced on the following pages.

Stick to the leader! Let "KENTONTOYS" show you the way to bigger turnovers!

There is a catalog and quotation awaiting your request to go forward to you.

BUY "KENTONTOYS"

NO. 620 BANK

Furnished with Combination Lock. Nickel Plated. Height 5¾ in. 1 in a box, 3 dozen in a case. Weight per case, 100 lbs.

50 Numbers of Banks and Safes

Every conceivable design and size, all shapes and finishes. A high grade line, one always saleable any time of year. These iron banks will outlast any other kind. Made with a regular dial combination lock.

NO. 145 BANK

Furnished with combination lock. Nickel Plated. Steel sides and back. Height, 6¼ in. Width 5½ in. Depth, 4½ in. 1 in a box. 2 dozen in a case. 100 lbs. per case

NO. 996

STOVE—GAS

Nickel Plated and Black Complete Double Gas Burners, Baking Oven Warming Oven—Utensils. Length, 12 inches. Height, 10 in. Width, 5 inches. 1 in a box. 1 dozen in a case. 200 lbs. per case.

80 Numbers of Stoves and Ranges

Every kind of design and size is in this line. Stoves that will actually burn gas—a regular gas burner by simply attaching rubber hose to gas attachment on wall.

NO. 960 STOVE

Nickel Plated. Length, 26½ inches. Height, 20 inches. Width, 12½ inches. Packed 1 in a case. 55 lbs. per case.

NO. 755 PISTOL

BLANK CARTRIDGE

Nickel Plated, Buff Finish. Length 5 inches. 1 dozen in a box. 3 gross in a case. 190 lbs. per case.

30 Numbers of Pistols

Both cap and blank cartridge pistols of best design. Beautiful nickel plated finish. Polished if you desire. A large range of sizes to select from and you are assured of getting the best made.

NO. 300 SAD IRON

Polished and Gold Bronzed. 1 dozen in a box. 2 gross in a case. 125 lbs. per case

15 Numbers of Sad Irons

From smallest to largest iron with Mrs. Potts detachable wood handle. Our larger irons are cast solid and are actually useful to iron out laces and fancy goods.

NO. 146 COAL WAGON

Two Horses. Painted and Decorated. Lines and Traces. 16½ inches long. 1 in a box. 2 dozen in a case. Weight per case 187 lbs.

120 Numbers of Wheel Toys

Road Carts, Express Wagons, Sulkys, Trolly Cars, Ice Wagons, Assortments, Transfer Wagons, Hansom Cabs, Coal Carts, Contractors Wagons, Farm Wagons, Drays, Surrys, Ox Carts, Mule Carts, Patrol Wagons, Every kind of wheel toys made. All sizes and all finishes at popular prices.

NO. 1820 AUTO FIRE ENGINE

Painted and Decorated. With Gong. 11½ inches Long. 1 in a box. 2 dozen in a case. Weight per case 200 lbs.

75 Numbers of Fire Toys

These are the favorites with the children. We make many varieties in auto and horse drawn fire toys which depict action.

NO. 846 PANAMA CONSTRUCTION TRAIN

Locomotive, tender, 3 dump cars Length, 36 in. Painted, 1 in a box, 2 dozen in a case. 180 lbs. per case.

90 Numbers of Trains

Passenger Trains, Coal Trains, Cattle Trains, Freight Trains, Combination Trains, Dump Trains. Any kind in operation is duplicated here. A complete range in sizes, number of cars and they are all priced right to sell.

THE KENTON HARDWARE CO., KENTON, OHIO

PAPER CAP PISTOLS

Representative Size of the Line of Cap Pistols

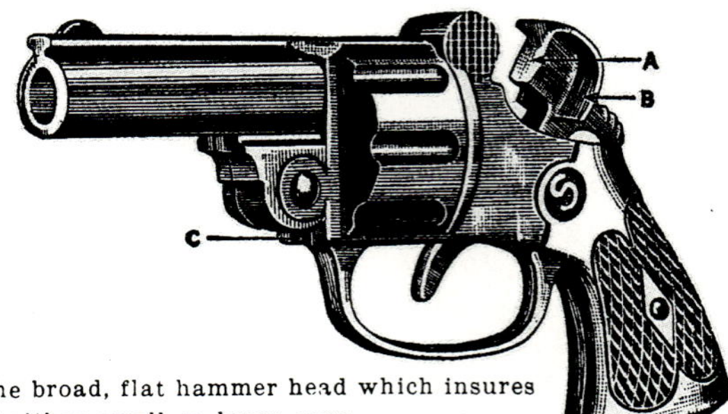

A—Illustrates the broad, flat hammer head which insures explosion of either small or large caps.

B—Shows the wide protecting Hammer flanges, preventing back fire and avoiding the possibility of sparks being thrown back on the hands or face of the operator.

C—Shows the forward vent or opening allowing exhaust of sparks and smoke from the exploded cap.

D—Shows the vent in cap receptacle allowing escape of sparks and smoke which passes down through the body of the pistol and out at "C."

E—Illustrates the new shoulder or offset to prevent sparks from splashing hands of operator. When cap is exploded sparks fly back on flange "B" where they are deflected downward to strike on shoulder "E" which entirely dissipates the force and danger from sparks striking the hand.

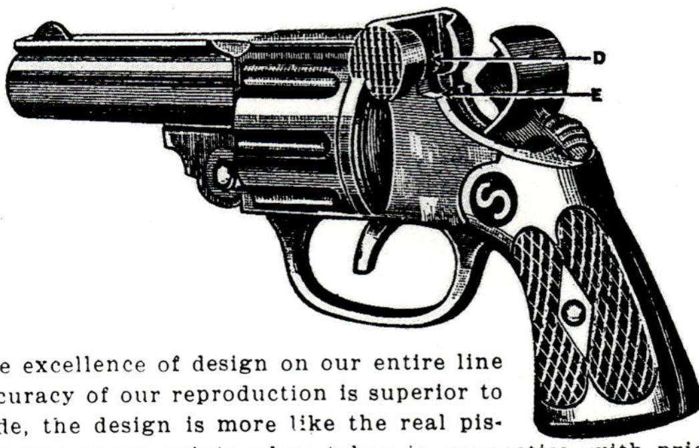

Note also the excellence of design on our entire line of Pistols, the accuracy of our reproduction is superior to other Pistols made, the design is more like the real pistol or revolver, these several points when taken in connection with pride in finish, care in inspection insures you of the utmost in value.

These improvements have not in any way added to the manufacturing cost. We are therefore able to offer you this new line of Pistols at the same price as asked for the older style cap pistols.

Design improvements as illustrated in pistol catalogs of the 1920s and 1930s. *Courtesy of Hardin County Historical Museums, Inc.*

Small cap pistols with small names. At left, *Dik*, nickel-plated, 5 1/8" long; at right, *Kido*, sample room toy, 5 3/8" long. *Courtesy of Hardin County Historical Museums, Inc.*

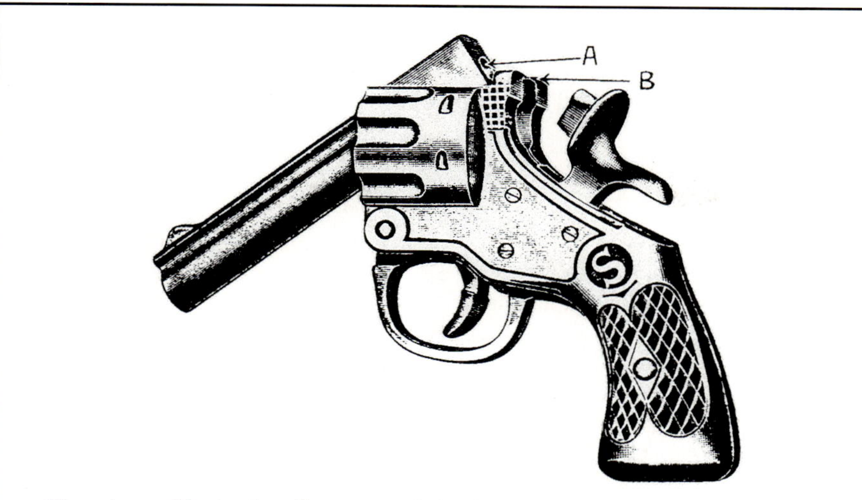

The above illustrates the new safety features which are now on all sizes and numbers of Blank Cartridge Pistols. The barrel "A" is countersunk only far enough to allow for insertion of .22 Calibre Blank Cartridge. If attempt is made to insert a .22 calibre ball cartridge, it will project out over the end of the barrel, because it is stopped at end of countersink--and the new shoulder "B" will prevent lowering the barrel to firing position.

Blank cartridge pistol safety features, as illustrated in pistol catalogs of the 1920s and 1930s.

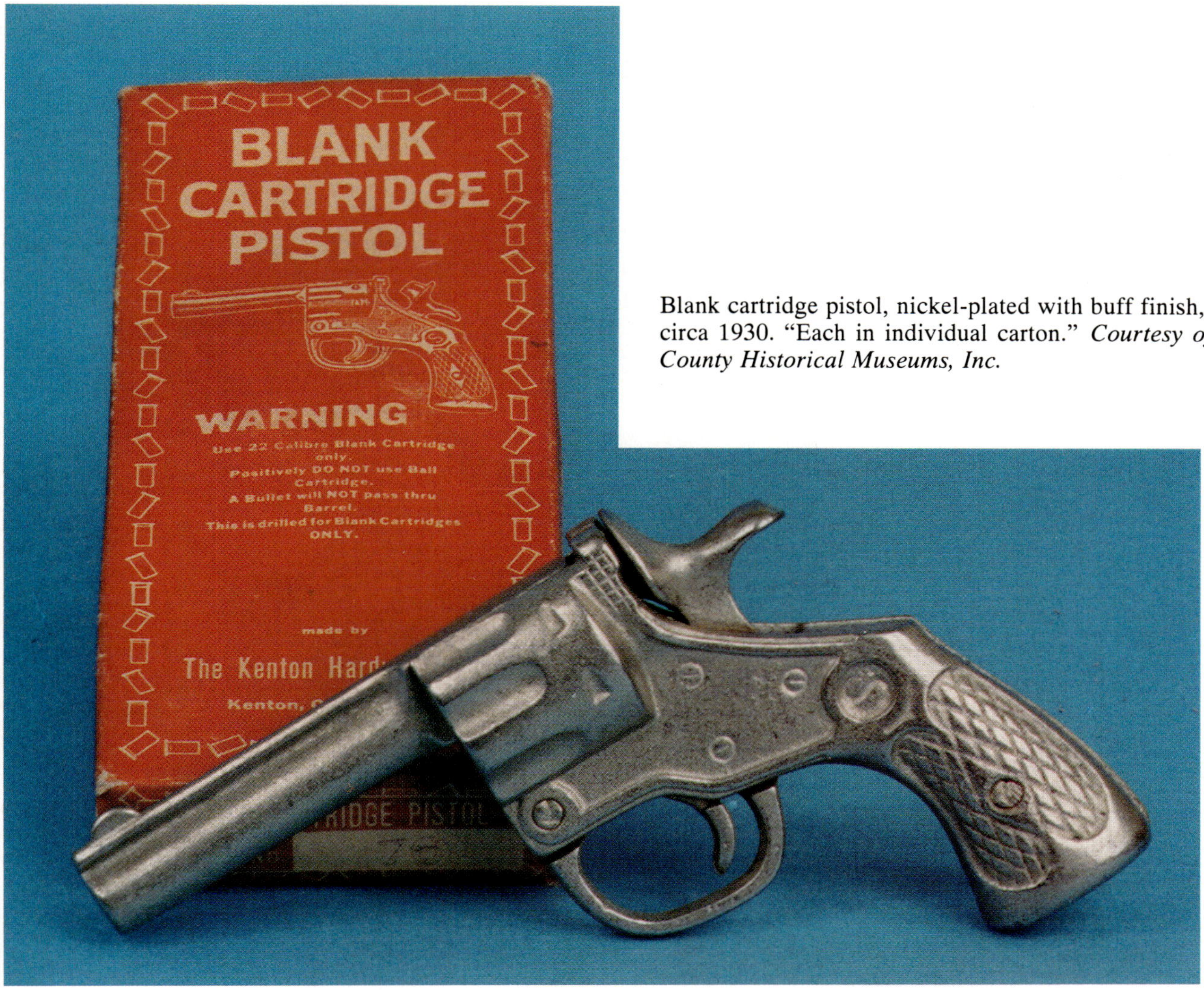

Blank cartridge pistol, nickel-plated with buff finish, 7" long, circa 1930. "Each in individual carton." *Courtesy of Hardin County Historical Museums, Inc.*

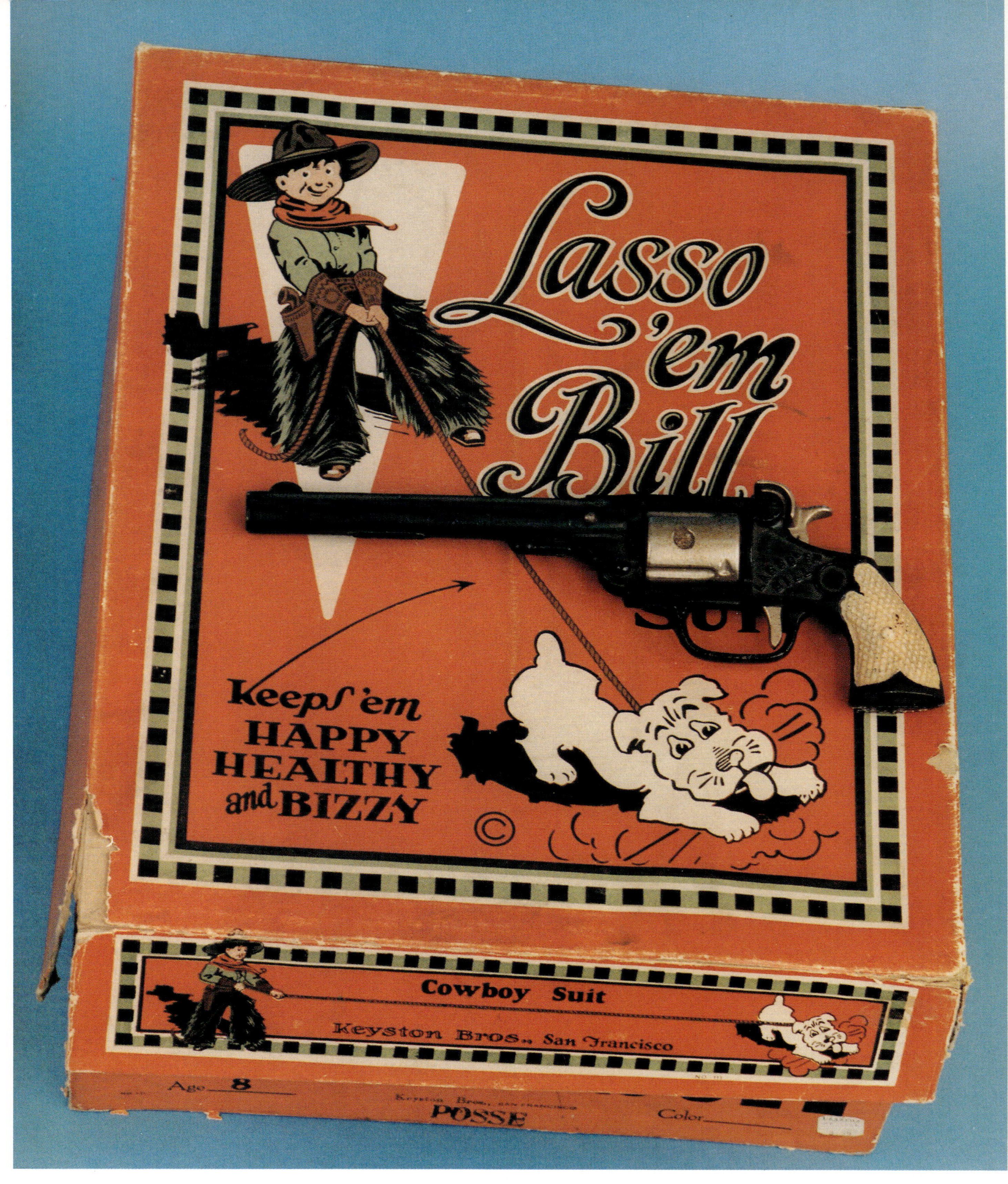

Lasso 'em Bill Cap Pistol, black, sample room toy, 9" long. The pistol is shown with a "Posse" cowboy suit box owned by Sally Wachalec Stevens, daughter of the plant superintendent. Lasso 'em Bill was a character licensed by the Keyston Brothers of San Francisco. *Courtesy of Hardin County Historical Museums, Inc.*

York Cap Pistol, black enamel, 7.5", nickel-plated, late 1920s. Sample room toy. *Courtesy of Hardin County Historical Museums, Inc.*

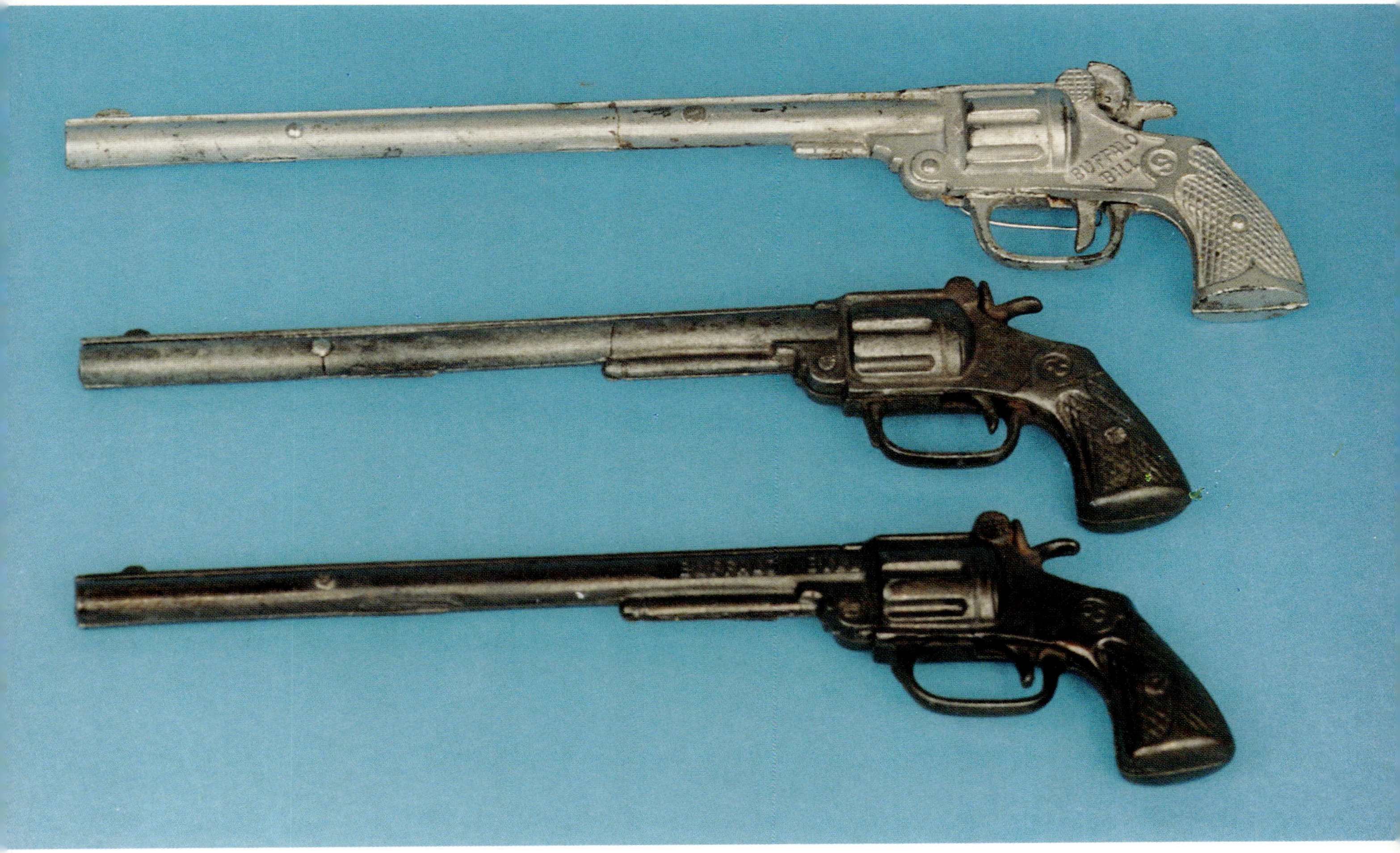

In its boom years, Kenton had no shortage of long-barreled cap pistols. Top: *Buffalo Bill*, nickel-plated, patented 1923, 13.5" long. Middle: *S & W*, copal varnish finish, 11.5" long. Bottom: *Buffalo Bill,* nickel-plated, 11.5" long.

Novelty Kenton Line Range, circa 1930, painted and nickel-plated, 8.75" x 11" x 6.25".

Kenton Line Ranges, circa 1910-1930, nickel-plated: left, 11" x 6.75" x 7"; right, 8" x 9" x 5".

Kent Gas Range, circa 1930, painted and nickel-plated, 8" x 7" x 4.5". *Courtesy of Hardin County Historical Museums, Inc.*

New Heatrola Bank, painted and nickel-plated, 4 5/8" x 2 5/8". Sample room toy with tag marked 1927. *Courtesy of Hardin County Historical Museums, Inc.*

Left: *Radio Bank*, painted and nickel-plated, 4.5" x 3 3/8". Sample room toy with original tag marked 1936. Right: *Radio Bank*, "New Combination. All Three Dials Operate." Painted and nickel-plated with original sample room tag marked 1927, 3.5" x 6 3/8". *Courtesy of Hardin County Historical Museums, Inc.*

Radio Bank, custom-finished by a Kenton employee with paint and gilt highlights, mid-1930s, 4.5" x 4". The dial of this Crosely model is tuned to 700 WLW-AM in Cincinnati.

"Carpet runners." Left: *Chicago, Rock Island, and Pacific* passenger car; right: *Chicago, Rock Island, and Pacific U.S. Mail* car, both circa 1925. Hitched to a locomotive and a tender, the train would measure approximately three feet in length.

July 29, 1924. **Des.** 65,319

L. S. BIXLER

TOY AUTO CIRCUS CALLIOPE

Filed March 25, 1924

WITNESSES

INVENTOR
L. S. BIXLER.

BY

ATTORNEYS

Overland Circus Hippopotamus Cage, 1924, 8" long. Another truck designed by L.S. Bixler. *Courtesy of Bill Bertoia Auctions.*

Design patent for L.S. Bixler's *Auto Circus Calliope*, 1924.

Lumber Truck with original planks, sample room toy with original tag marked 1930, 11.75" long. *Courtesy of Hardin County Historical Museums, Inc.*

Lift Dump Truck with hinged, self-locking gate, 6.5" long. Sample room toy with original tag, circa 1930. *Courtesy of Hardin County Historical Museums, Inc.*

Left: *Coal Truck*, sample room toy with original tags, 10 3/8" long, early 1930s. Right: *Auto Dump Truck*, sample room toy, 6.5" long. *Both courtesy of Hardin County Historical Museums, Inc.*

Left: *Sprinkler Truck* with original sample room tag, marked 1927, 5 5/8" long. Right: *Gas Truck* with original sample room tag, marked 1936, 7.25" long. *Both courtesy of Hardin County Historical Museums, Inc.*

Gas Truck (3.75" long) with short (4.5" long) and extended (6" long) tank trailer. Sample room toy with original tag marked 1931. *Courtesy of Hardin County Historical Museums, Inc.*

Left: *Sedan*, "Latest model with Detached Wheel," circa 1925, 5.5" long. Right: *New Model Coupe*, "Detachable Wheel," circa 1925, 5.5" long.

Sport Racer, circa 1925, 10.25" long.

Sedan ("Red Top Cab"), 8.25" long, circa 1926. *Courtesy of Bill Bertoia Auctions.*

Racers, sample room toy with original tags. Left: 4 3/8" long, marked 1936. Right: 7.5" long, marked 1934. *Courtesy of Hardin County Historical Museums, Inc.*

Racers, sample room with original tags. Left: 7.5" long, marked 1934. Right: 7 3/8" long, marked 1934. *Courtesy of Hardin County Historical Museums, Inc.*

Inter-City Bus Coach, late 1920s, 13.25" long. *Courtesy of Bill Bertoia Auctions.*

City Bus, sample room toy with original tag marked 1926, 6" long. Sold without figures. *Courtesy of Hardin County Historical Museums, Inc.*

City Buses, sample room toys with original tags. Left: 12" long, marked 1924. Right: 10" long, marked 1926. *Courtesy of Hardin County Historical Museums, Inc.*

Pickwick Nite Coach, sample room toy with original tag, 1928, 7.5" long. *Courtesy of Hardin County Historical Museums, Inc.*

Pickwick Nite Coach, nickel-plated grill and exhaust pipe. The 1932 catalog described this as a "new model (which) conforms to the latest design of the Pickwick Nite Coach now in use." *Courtesy of Bill Bertoia Auctions*

Express Wagon, overall length 7 3/8", made by (and marked) National, though not identified as such in catalogs.

Jaeger Concrete Mixers: left, 10" long; right, 7.25" long. "At first glance, this would appear to be a large Jaeger Concrete Mixer. However, it is a Kentontoy," according to 1929-1932 catalogs. *Courtesy of Hardin County Historical Museums, Inc.*

Jaeger Concrete Mixers, with rubber wheels, sample room toys, circa 1940. Left, 8.5" long; right, 7.25" long. *Courtesy of Hardin County Historical Museums, Inc.*

Jaeger Drum Type Mixers, sample room toys with original tags, early 1930s. Left, 9.25" long; right, 7" long. "When this realistic little truck is pulled on the floor, the drum automatically revolves, mixing contents such as sand or small stones," said 1931-1932 catalogs. *Courtesy of Hardin County Historical Museums, Inc.*

Pony Blimp, circa 1930, 6.75" long. This toy was inspired by Goodyear's "aerial yacht."

Los Angeles, circa 1930, 8" long. "The air-minded lad will want one or all of the ships of the air so closely modeled after the real thing," said 1929-1932 catalogs.

Lucky (as in "Lucky Lindy") with nickel-plated propellor and wheels. Early 1930s, 6 3/8" long.

The *Air Mail*, like Charles Lindbergh's Ryan Monoplane upon which it was based, lacked a windshield. ("Lucky Lindy" used a periscope.)

Air Mail Planes, 9 7/8" long, the largest version of this toy.

Steam generating plant in Kenton factory, circa 1930. *Courtesy of Hardin County Historical Museums, Inc.*

HORIZONTAL ENGINES

By turning the crank, all working parts of the Engine are put into operation. By contact with pulley on crank shaft with round rubber belt, a small pulley operates the governor by means of gears. The eccentric pulley on crank shaft operates rocker arm which in turn moves the valves on steam chest in realistic manner by means of connecting wires. On the opposite side, the rotating crank head operates connecting rod which moves the Piston back and forth in realistic manner.

Catalog No.	Length	Width	Height	Pieces per Box	Dozen per Case	per Case Weight
500	8"	5¼"	5"	1	2	90
501	10½"	5⅝"	5½"	1	1	75

VERTICAL ENGINES

Here is another model that is a true copy of a large power producer. All parts work from crank shaft, except governor which operates by belt from pulley on crank shaft. Piston works in vertical position, operating valve on cylinder head.

Cat. No.	Length	Width	Height
502	6⅜"	3⅛"	8½"
503	7¾"	3⅜	9⅝"

No. 502—1 in a box. 2 dozen in a case. Wt. of case, 75 lbs

No. 503—1 in a box. 1 dozen in a case. Wt. of case, 50 lbs.

THE REAL THING IN EVERYTHING BUT SIZE 

17

Page from Cast Iron Toy Catalog of 1932.

Horizontal engine at Kenton factory, circa 1930. *Courtesy of Hardin County Historical Museums, Inc.*

The *Morgan Crane*, early 1930s, 16" x 12.25" x 12". "Three way movement of the load is governed by two cranks and the wheel on the control car. The lever at the top allows the load to drop immediately. Only four bolts and nuts are necessary to erect this colorful toy," explained the 1932 catalog.

Buckeye Ditcher, sample room toy with original tags marked 1930; nickel-plated mechanisms. Left: 13.75" x 2.5" x 5". Right: 8 7/8" x 2.5" x 5". *Courtesy of Hardin County Historical Museums, Inc.*

Power Shovel, early 1930s, 5.5" long. Sample room toy with original tag. Part of a line of road building equipment. *Courtesy of Hardin County Historical Museums, Inc.*

FAIRFIE

Fairfield Loader, early 1930s; left, 12.75" long; right, 9.75" long. Promoted as "A Joy for Every Boy," and the perfect complement to Kenton's toy trucks and concrete mixers.

Tractor, sample room toy with original tag, marked 1928, 6.5" long. Inspired by a Ford model, this design (in four sizes) was Kenton's sole entry in the tractor line. *Courtesy of Hardin County Historical Museums, Inc.*

Beer Truck, 1933 (the year Prohibition was repealed), 15.5" long. Originally supplied with beer kegs. "The old time beer truck is back and Kenton again leads with an interesting model of a horse drawn load," stated the 1933 catalog supplement. The return of miniature wagons was a milestone in cast iron toy history and anticipated the "nostalgic" line of the next decade. *Courtesy of Bill Bertoia Auctions.*

Dray Wagon, sample room toy with original tag marked 1933, 15.5" long. *Courtesy of Hardin County Historical Museums, Inc.*

(Speed) *Truck Trailer*s, sample room toys with original tags. Left, top: 10.5" long, marked 1929. Left, bottom: 8" long, marked 1937. Right: 8" long, marked 1936. *Courtesy of Hardin County Historical Museums, Inc.*

Ice Trucks, sample room toy with original tags marked 1926; left, 6" long; right, 7.75" long. *Courtesy of Hardin County Historical Museums, Inc.*

Fire Engine Truck (pumper) with gong, sample room toy, circa 1930, 10.75" long.

Hook & Ladder, sample room toy with original tag, circa 1925, 14 1/8" long.

Firemen's Patrol, sample room toy with original tag, late 1920s, 7.5" long. *Courtesy of Hardin County Historical Museums, Inc.*

Sample room toy automobiles with original tags. Left: *Chief*, marked 1930, 5.5" long. Right: *Fire Chief*, marked 1933, 6 1/8" long. *Courtesy of Hardin County Historical Museums, Inc.*

The cautiously optimistic vice president, Willard Bixler, at work. A portrait of his father, L.S., hangs at upper right.

The packing room, circa 1930.

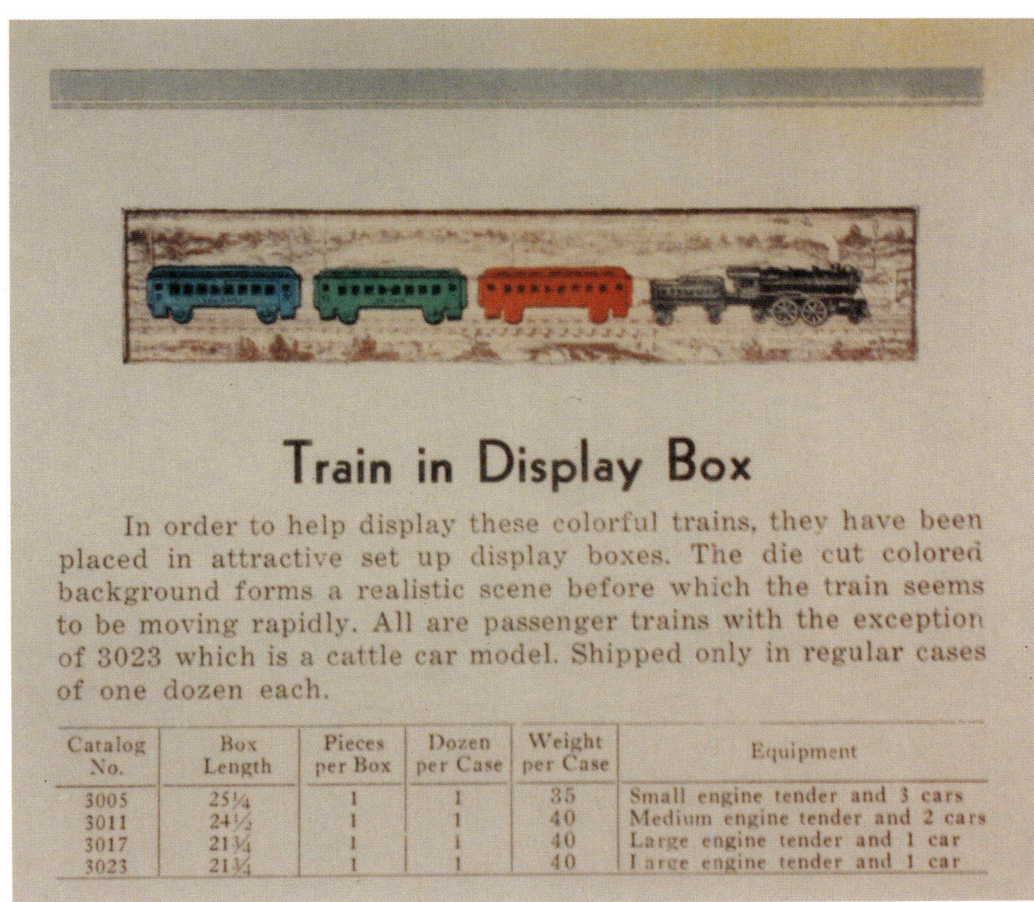

Train in Display Box

In order to help display these colorful trains, they have been placed in attractive set up display boxes. The die cut colored background forms a realistic scene before which the train seems to be moving rapidly. All are passenger trains with the exception of 3023 which is a cattle car model. Shipped only in regular cases of one dozen each.

Catalog No.	Box Length	Pieces per Box	Dozen per Case	Weight per Case	Equipment
3005	25¼	1	1	35	Small engine tender and 3 cars
3011	24½	1	1	40	Medium engine tender and 2 cars
3017	21¾	1	1	40	Large engine tender and 1 car
3023	21¾	1	1	40	Large engine tender and 1 car

In the first years of the Great Depression, Kenton followed the leads of Arcade and other rivals in marketing eye-catching display boxes.

Boxes for individual toys, circa 1930-1950.

Airport Assortment as illustrated in the catalog.

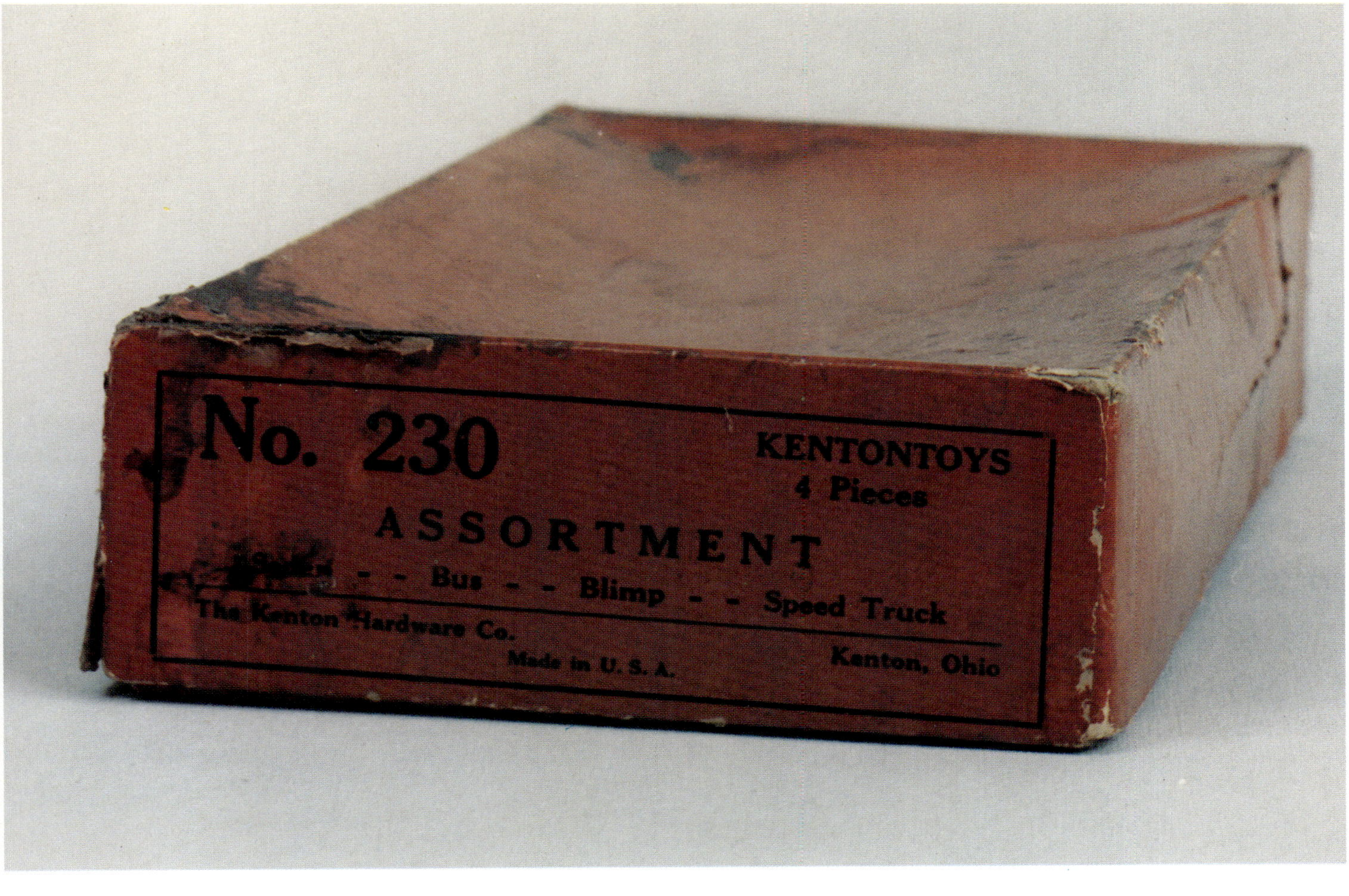

End view of rare carton for *Airport Assortment*, 14.25" x 7.25" x 2.25", circa 1930.

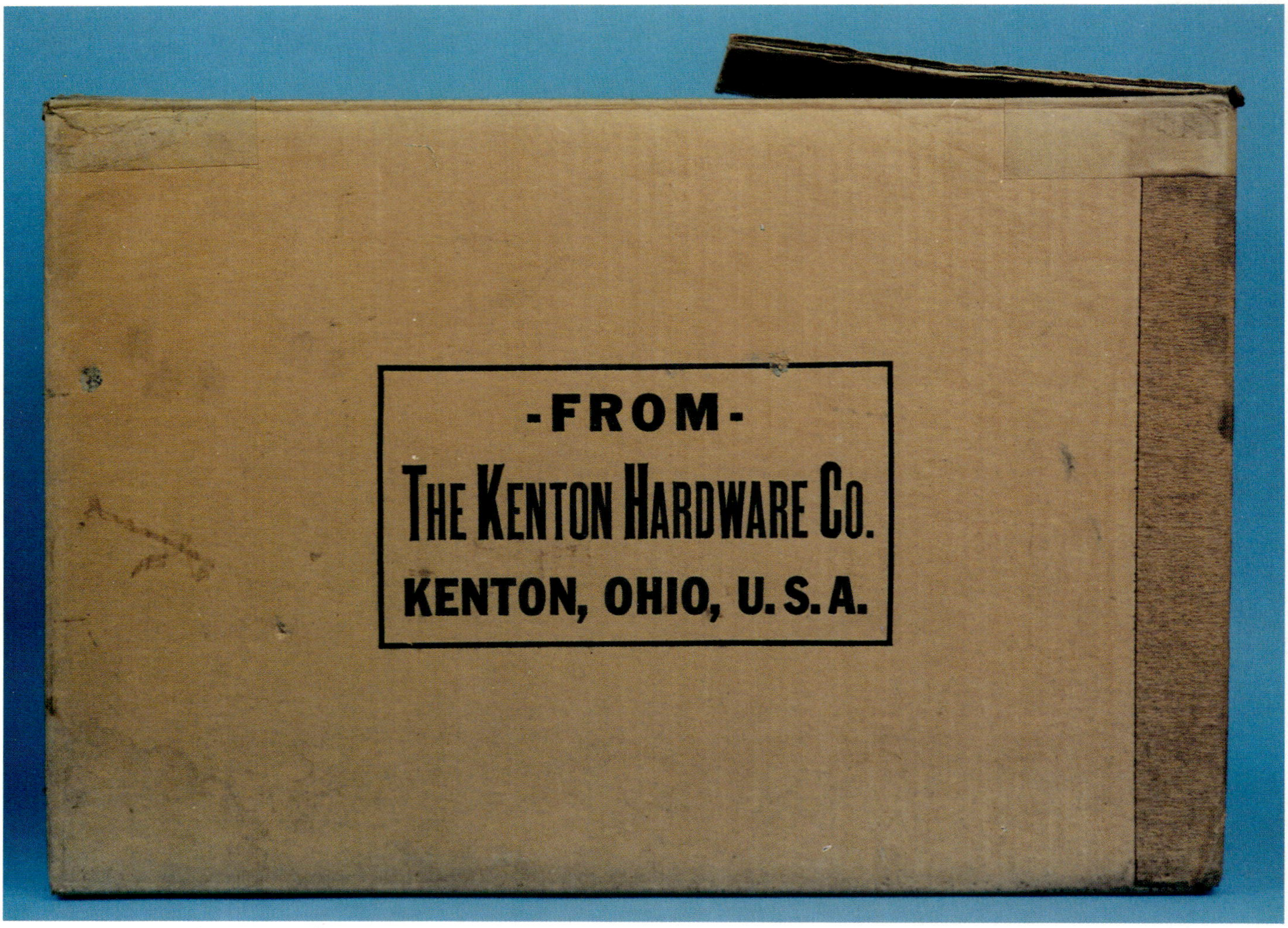

Kenton shipping carton, circa 1930 to circa 1950. *Courtesy of Hardin County Historical Museums, Inc.*

Bull Dog Cap Pistol, deluxe version with coppered trim and jewelled grip. Sample room toy collection, circa 1930, 6.75" long. *Courtesy of Hardin County Historical Museums, Inc.*

Sample room toy cap and cartridge pistols (from top, clockwise) *S&W*-type, nickel-plated with jeweled grip, circa 1925, 6.75" long; *S&W*, nickel-plated, circa 1925, 6.25" long; *Dixie*, nickel-plated with jeweled and painted grip, circa 1937, 6.25" long; *Westo*, nickel-plated with jeweled and painted grip, circa 1936, 7" long; and *Cartridge Pistol*, nickel-plated with buff finish, circa 1925, 5" long. *Courtesy of Hardin County Historical Museums, Inc.*

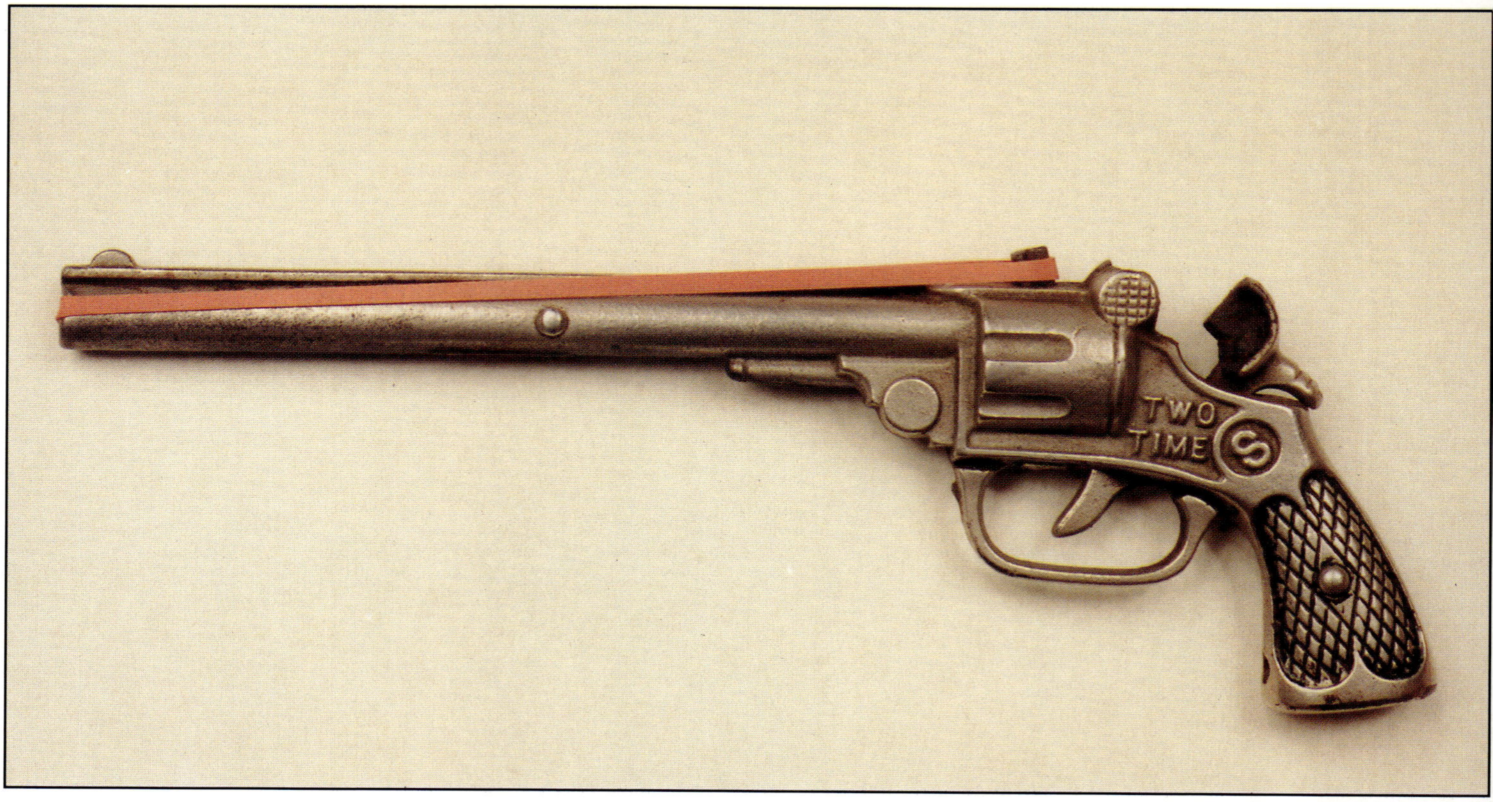

Two Time Cap Pistol & Rubber Band Shooter, buffed nickel finish, circa 1930, 9.75" long. "Will shoot rubber bands only, or paper caps only, or both simultaneously," explained pistol catalogs.

Circa 1933 letterhead incorporating the new Kenton logo, the National Recovery Administration "Blue Eagle" (showing that Kenton had endorsed the Code of Fair Competition for the Toy and Playthings Industry), and Toy Manufacturers of the U.S.A. indicia.

Quick on the Draw . . . But Not Quick Enough

Kenton relied on its pistol line for relief. The *Two Time Target Pistol*, which fired a cap and a rubber band simultaneously, was touted as a "profit builder," and Kenton claimed it was hundreds of gross behind in fulfilling orders for the product. The *Riot Gun*, inspired by crank-operated pistols "used by police all over the country," was expected to be gobbled up by aspiring G-men, blasting away the competition. However, it remained for another Kenton pistol to create a riot in the toy marketplace.

Harold "Buddy" Spencer, jack-of-all-trades for the Kenton Hardware Company, approached master pattern-maker Joe Solomon with an idea for a toy gun. The youthful Spencer had been impressed by a carriage-mounted anti-aircraft machine gun he had seen in his military service. He believed youngsters would love playing with a rapid cap-firing blaster. Solomon crafted an intricate prototype, complete with a crank and protective breech.

Unfortunately for Spencer and Solomon, the Grey Iron Casting Company of Mount Joy, Pennsylvania, was quicker on the draw. In 1928 they marketed a *"Knockout" Rapid-Fire Anti-Aircraft and Machine Gun,* and the Kenton design never went into production.

Today the unique prototype stands silently in a showcase of cast iron toys in the Sullivan-Johnson Museum in Kenton.

An advertisement from 1928.

Anti-aircraft/machine gun prototype designed by Harold "Buddy" Spencer and Joe Solomon. Brass pattern toy, 7.5" x 4 5/8". *Courtesy of Hardin County Historical Museums, Inc.*

Chapter 9
Great Guns!

In 1937, Gene Autry and the Kenton Hardware Company met going in opposite directions.

The Singing Cowboy was a thirty-year-old shooting star on a dramatic rise. The young telegrapher and telephonist went from honkytonk to radio station to recording studio to Hollywood, all in the course of a decade. As Public Cowboy No. 1, he embodied family values and the virtues of the American way of life. In formulaic operettas, Autry and his trusted steed Champion confronted modern problems in frontier settings.

Autry proved to his admirers that even in the Great Depression good would triumph over evil, that the Protestant work ethic would prevail over crooked business, and the common folk would win in the end.

Kenton, on the other hand, was on the wane. The cast iron toy trade was hit hard by the Depression, and Kenton and its competition suffered substantial drops in orders. The 100-plus page catalog of the heyday of the 1920s dwindled to little more than a leaflet by 1933.

The Chicago's World's Fair of 1933-34 offered the opportunity for Kenton to merchandise the progressive–though sadly slow-moving–*Tour Bus* and *Century of Progress Train* (inspired by the *City of Salina* streamliner). A new U.S. president inspired a *New Deal Bank*. At the same time the New Deal itself, with incentives provided by a Code of Fair Competition for the Toy and Playthings Industry and by the National Recovery Administration, promised a brighter future. However, continued sluggish sales dampened spirits. Advertising tailed off, production slipped to an estimated forty percent of previous banner years, and workers, instead of being laid off, were limited to two working days per week.

Undoubtedly many looked for the unsinkable L.S. Bixler to rise to the occasion as Kenton slid into the Great Depression. But this time around, it would be the son, not the father, who would have the ace in the hole.

Young Bixler Comes Through

Willard R. Bixler had maintained a long-standing interest in cap pistols. Correspondence from 1924 indicates his exhaustive and eventually futile efforts to patent a pistol capable of firing standard blank cartridges of any and all calibers. He did enjoy success with a subsequent invention, a 1937 patent for the improvement of a trigger-hammer cap feed arrangement.

Carl Drumm, the "Rambling Reporter" for Kenton's *News and Republican*, revealed in 1938 that for well over a year an idea had been germinating within Willard Bixler's mind. He wanted to develop a quality cap pistol that looked like a real gun.

> The idea involved the manufacture of a deluxe cap pistol capable of selling at twice as much as any ordinary cap pistol ever brought before. Design of such a pistol was easy. But collecting half a dollar each was something else again. There had to be some built-in attraction not possessed by other pistols.
>
> The idea grew and developed–until one day it blossomed forth in sudden inspiration. Why not design the toy after the gun used by some famous movie star?

Willard Bixler, perhaps influenced by trade journals which had prophesied a boom in cowboy toys, instituted a search for the most popular Western star in the business. A business associate, Edward Gruskin, suggested Gene Autry. Autry had received more fan mail than Clark Gable, Myrna Loy, Spencer Tracy, or (according to Drumm) "any other star known to the Los Angeles post office."

Vice president Bixler approached New York whiz kid Mitchell J. Hamilburg, the merchandising representative of Autry, Edgar Bergen (and Charlie McCarthy), Deanna Durbin, and other personalities. Hamilburg saw great promise in Bixler's idea, and told a journalist, "There is no doubt whatsoever that 'characters' have been the biggest stimulant that the toy business has ever had."

Bixler himself told reporter Carl Drumm,

> . . . youngsters would be in Seventh Heaven if they could pack on their hip a replica of the gun. . . . The cowboy [Autry] sent to Kenton the very same, pearl-handled gun he [used] in his rip-rarin' sagas of the West. Patterns and molds were made from the weapon, and after the expenditure of $10,000, the first cap pistol was ready for the market.

The *Gene Autry Toy Pistol* was unveiled in the December 1937 issue of *Playthings*. A full-page advertisement waxed, "It sells on sight. Boys demand it! Dads can't resist it! Couple this irresistible quality with the unmatched publicity value in the name of Gene Autry, Republic Pictures' foremost star . . . and you have a selling combination that can't be beat." The promotion continued,

> Here it is, the most beautifully finished repeating toy cap pistol on the market–made with gun metal finish and separate models of imitation pearl. Gene Autry's signature appears as part of the pistol. The construction is safe and sturdy–it's made to last. This gun is bigger than any Cap Pistol offered to the trade, and it is accurately modeled from Gene Autry's own gun. Positively the finest toy pistol value ever offered.

The toy managers held their collective breaths: would the thing sell? Preliminary indications from jobbers suggested that the high price tag would scare away potential customers.

The response must have surprised even Willard Bixler himself.

> The toys began to sell. They sold from coast to coast. They sold in Cleveland, where there [was] a strict ordinance against cap pistols. They sold in the Latin American countries. . . . They sold in Europe, especially in London and Scotland. . . . The capacity of the local plant was severely strained by the rush of orders that followed introduction of the pistol. The force was doubled, and still the plant was unable to keep up with demand . . . so the local plant operated day and night, and some work was done in other foundries.

According to Willard Bixler, one million *Gene Autry Cap Pistols* were manufactured between February 10 and the first week of August 1938. By December they were calling it "the sensation of the decade." Through jobbing and chain stores, they enjoyed the widest distribution of any Kentontoy ever. They were shipped to many foreign countries, and, as one contemporary journalist noted, their appeal resulted in reorders. Sales were especially brisk in Latin America.

By the fall of 1939, the *Saturday Evening Post* reported that the total number of *Gene Autry Cap Pistols* manufactured had reached two million. *Post* feature writer Alva Johnston observed that the factory was still operating day and night, adding, "The Singing Cowboy's fan mail has been increased by letters complaining that he has turned every day into the Fourth of July."

However, complaints were drowned out by acclamation. In almost miraculous fashion, the fortunes of the Kenton Hardware Company had been reversed. Gene Autry, who had himself received $12,000 in royalties during the first four months of cap pistol production, was hailed by Kentonite and non-Kentonite alike as an "economic savior." The singing cowboy himself referred to his cap pistol as the toy pistol that "saved an entire town."

Shortly after the sale of the millionth cap pistol, the people of Kenton would have an opportunity to thank Gene Autry in person. A strike for higher salaries at Republic Studios allowed the Western star and a "troupe," consisting of his wonder horse Champion, comic Frankie Marvin, the musical Cass County Boys, and "extras," to tour the Midwestern and Eastern states. Manager George Goodale had arranged for Kenton to be the only Ohio stopover between engagements in Pittsburgh, Pennsylvania, and Flint, Michigan. The appearance, to paraphrase a contemporary Kenton newspaper account, would be as much to see how his cap pistols were made as to meet his numerous admirers in Hardin County.

Gene Autry Arrives

Gene Autry and company arrived in Kenton on the morning of Monday, August 8, 1938. The arrival had been heralded by local papers; one reporter predicted that "practically every hero-worshipping boy and girl in the district" would turn out to meet him. That they did, mobbing him at every turn.

A late morning tour of the plant and its various departments was punctuated by intrusions of children wriggling through windows, and foremen wisely shut down operations.

According to one eyewitness account, a persistent lad pushed his way through a cordon, reached the Singing Cowboy, and tugged on the star's monogrammed sleeve. The child asked, "Gene, can you give me a signed picture of you and Champ?" Gene, waving off the converging adults, answered, "Gee, I'm sorry, son. I'm plum out of photos. But I'll tell you what–I'll be sure to send you one just as soon as I get back to Hollywood."

Without a moment's hesitation, the boy shot back, "Like he** you will!"

The tour took a surreal turn when Champion, riding a service elevator to the second floor, joined his master on the factory tour. Gene Autry and the Kenton Hardware Company workers paused to pose for photographs inside and outside the plant. As the visit drew to a close, one employee noticed that Autry's light-colored cowboy duds had taken on a gray cast from the factory's abundant dust and dirt, much of it transferred by children's embraces.

What was to be a relaxing dining experience at the Kenton home of Willard and Helen Bixler turned out to be yet another mob scene, as youngsters surrounded the North Detroit Street residence, jostled for space on the front porch, and peered in windows.

A "Gene Autry Day" was planned for Kenton's Central School playgrounds. The program included a rodeo, track and field events, a softball game, and a personal appearance by the cowboy star. The *Kenton Daily Democrat* reported that fifty ice cream cones and three passes to the Kenton Theatre would be awarded to the winners of the various events.

Earlier in the day, a line began to form outside the Kenton Theatre in anticipation of the first of five performances at 3:30 p.m. Theater manager G.H. Foster had splashed large display ads in the local papers, calling the live shows "AN EVENT EXTRAORDINARY . . . Adults 26¢, Children 16¢." The ads emphasized that Gene Autry would perform in person with movie comic Frankie Marvin and a company of entertainers. Further enticement was not really required, but nonetheless Foster added a screening of Allan Lane and "The Duke Comes Back."

The lines snaked around the block, and youngsters and their parents filled the sidewalk from one side to the other by curtain time. A traffic jam the likes of which had never been witnessed in Kenton brought cars to a standstill on the outskirts of town. By the end of the fifth performance, an estimated 4,500 plus would jam the theater, breaking local attendance records.

Kenton's senior citizens look back on August 8, 1938, as Kenton's red-letter day. They warmly recall the electricity of anticipation, the thunderous applause which greeted the Singing Cowboy, the official reception by city fathers who thanked Gene for "saving their city," the fancy "strummin' and yodellin'," and the hoof prints–among other things–that Champion left on the stage.

Gene Autry recalled that after his last bow in the first show, the children refused to leave their seats. He returned to the stage and offered cap pistols to the first twenty-five youngsters out the door. He might as well have cried "fire!" Autry's poor manager, George Goodale, was trampled in the stampede.

As midnight approached, Autry and his cowboy troupe quickly packed and hit the road, bound for Flint, Michigan.

Production quickly resumed at the factory. As the "Rambling Reporter" Carl Drumm told his readers, "In the toy business you have to sell when there is a demand, for the juvenile fancy is notorious for its fickle nature . . . You have to sell while the market is clamoring for your goods."

To the amazement of not a few traditionalists in the toy world, Gene Autry had dominated the cap pistol industry by the end of 1938. The *Gene Autry Repeating Cap Pistol* established itself as the mainstay of the Kenton Hardware Company.

Kenton's Cap Pistol Line

The company was quick to build on the success of its sensation. "The immense popularity of this pistol in 1938 clearly demonstrates that it will be in 1939, the item around which a cap pistol line should be developed," Kenton revealed within a year of the toy's introduction. The '39 Gene Autrys were offered with imitation pearl and "new brilliant" red molded plastic grips branded with the figure of a horse. Improved two-color cartons would be expected to make the pistol "even a greater sales producer."

Kenton added other new and improved models to its pistol line, including the *C-Boy*, *Western*, *Police Chief*, and *Police Chief, Jr.* A later model, the *Dude*, "right off the Dude Ranch with Plenty of Class," enjoyed brisk sales. An advertisement previewed a *Gene Autry Play Gun* (dummy) "identical in every respect to the regular *Gene Autry Cap Pistol*s, except that it is impossible to explode or fire a cap with this gun." Kenton recognized "tremendous sales possibilities in territories where regular caps [were] prohibited by law."

Kenton rode their pistol like the Singing Cowboy rode Champion. A reproduction of the *Gene Autry* box became a standard in Kenton advertising. A silhouette of the six-shooter, symbolic of "a line of real quality," graced the covers of Kenton's pistol catalogs.

Hubley and other competitors were momentarily left in the dust. The Kenton pistol invoked a wave of hurriedly-produced imitations, but Kenton's version of the Autry Colt .45 won the West . . . as well as the North, South, and East.

A Departure for Kenton

In early 1939, Gene Autry, Mitchell Hamilburg, and L.A. Carll (who had been appointed president of both Kingsbury and Kenton in 1937) held a conference on the Pacific coast. They announced that Kenton would produce *Gene Autry's Bandit Trail* board game. The toy world was taken by surprise; the new plaything marked a dramatic departure from Kenton's long line of cast iron toys. With not a little fanfare, *Bandit Trail* made its debut at the American Toy Fair in New York on April 17. The game was extravagant, even opulent, by Kenton standards: the box was a three-color "grabber," the heavy imitation leather playing board featured a spectrum of seven colors, and thirteen lead markers were crafted in shapes of "characters of new design." It was touted as a "truly new game of exceptional counter and eye appeal . . . A SURE HIT–At $1.00 Retail."

The "sure hit" turned out to be a sure miss. *Gene Autry's Bandit Trail* was advertised for a year, then faded into oblivion.

Fortunately for Kenton, the Gene Autry pistol remained "the youngsters' first choice." The company's stock-in-trade afforded an opportunity, if not luxury, to launch another trial balloon.

Kenton's Nostalgia Line

In early 1941, Kenton announced to the trade, "The Horse Drawn Vehicle of Yesterday has Joined the Streamlined Equipment of Today IN OUR DISTINCTIVE AND UNUSUAL IRON TOYS." Although the "nostalgia line" might have appeared as an anachronism–and another doomed experiment–to many in the toy business, Kenton's strategy was well-reasoned and inspired. A wave of nostalgia had swept the country as Americans retreated to "the good old days" as a refuge from the grim realities of modern, global war. Romantic historical dramas became favorites among moviegoers. "Early American" became all the rage in domestic decor, and several magazine writers encouraged the use of cast iron toys of bygone days, "when the world was young and marvelous to behold," as decorative accents.

The horse-drawn line marked a radical departure from Kenton's "up-to-the-minute," "most modern design" philosophy. The change had been foreshadowed by a *Beer Truck* (released in conjunction with the repeal of Prohibition in 1933) and related work wagons. The *Beer Truck* et al had also anticipated a shift in marketing; nostalgic wheel toys were designed as conversation pieces for adults, and perhaps secondarily as playthings for their children. The visionary Bixlers and company looked to tap into the gestational toy collectors' market.

Kenton's 1941 catalog was a rather odd concoction of the old and the somewhat new. The popular *Jaeger Mixer* with balloon tires was spotlighted on the cover. Within the publication could be found jaunty models of the horse-drawn *Circus Cage* (reminiscent of Hubley's *Royal Circus* vehicles, if not Arcade's *Circus Wagon,* and *Big Six Circus & Wild West Wagon*), as well as the *Circus Calliope*, *Fire Engine*, and *Milk*, *Dray*, *Dump*, and *Stake* wagons. Alongside these nostalgic playthings were found the modern horseless *Fire Engine*, *Hook and Ladder*, *Fifth Ave. Bus*, and *Buckeye Ditcher*. Two sizes of the *National Combination Safe* and three sizes of the *Royal Range* represented the ends of Kenton's long and illustrious bank and stove lines. Together these were toys that were guaranteed to "sell–and repeat well."

World War II

Global war and national defense measures had significant adverse effects on the domestic toy trade. Exports plummeted, prices of raw materials rose, labor and parts shortages occurred, and toymakers were unable to guarantee prices or deliveries. Arcade and other cast iron toymakers announced that they would eliminate models and accept orders for "only stock in hand." Kenton officials looked to convert over to the production of war materials.

Little did Kenton executives realize that America's entry into World War II would prevent the company from living up to its earlier guarantee of toy sales and repeat sales. In March of 1942, vice president and general manager Willard R. Bixler told an interviewer,

> Many people think that all that is necessary to do is to go out and grab a Defense order, but it isn't as simple as that. In our plant we have no machinery of the type that can be used to fashion war materials. . . . In our case we are equipped to make light gray iron castings only, and we are not equipped to make what is known as heavy or machine castings, nor do we have the technical knowledge for such type of work. Our whole set-up is for the manufacturing and handling of the light kind of castings only, and because of this our manufacturing capacity is limited. In the present picture there is very little demand for light castings.

Bixler reported that his plant had been busy with regular toy orders. He rationalized, "Toys are an educational necessity . . . this is the period in which children need playthings the most." He added that four hundred Hardin County residents and "every merchant in this community" were dependent on his operation, and that he and his fellow executives were "doing everything possible to protect it."

Bixler's appeals and efforts were in vain. The War Production Board's limitation order L-81, made effective April 1, 1942, cut off production of metal toys. The factory closed down "for the duration," no war work having been found.

Roosevelt (New Deal) Bank, introduced in 1933, with custom gold-bronzed finish applied by a Kenton Hardware employee, 5" x 2 5/8". "An inspiring toy and a beautiful mantel-piece for any room," according to the 1933 catalog supplement.

A desk ornament/paperweight adapted from the *Roosevelt Bank* by another Kenton employee. *Courtesy of Hardin County Historical Museums, Inc.*

F.D.R. in profile, with a standard painted finish.

Frances Wachalec, the superintendent's wife, provided her sugar bowl to be used as the model for Kenton's *Duck Bank*. On the right is a patinated pattern toy. *Courtesy of Hardin County Historical Museums, Inc.*

Left: *Duck Bank*, 4" x 5". Right: *Duck Banks* used by the Wachalecs as caps for the fence surrounding their home. Note the hole drilled at the base. *Courtesy of Hardin County Historical Museums, Inc.*

Left: *Land-on Roosevelt* novelty, 1936, 5" long. Adapted from the *Cairo Express* wheel toy of the previous generation, this is a clever word play on the presidential candidates' names. The product was perhaps conceived by L.S. Bixler, a prominent Republican. *Courtesy of Hardin County Historical Museums, Inc.* Right: elephant novelty, copper finish, 5" long. Another knickknack adapted from the *Cairo Express.*

Sample room toy pattern toys with original tags marked 1937. Left: *Mule* (Donkey) *Bank*, 6". Right: *Elephant Bank*, 5". The corresponding finish suggests these were a political pair. *Courtesy of Hardin County Historical Museums, Inc.*

Kenton made cleats which were used as clothesline fasteners, blind and curtain tiebacks, etc. Donkey, 2" x 3 1/8"; Elephant, 2 1/8" x 3". *Courtesy of Hardin County Historical Museums, Inc.*

Tabletop Stove, enamel and nickel-plated, circa 1936, 4" x 2 1/8".

Wrecker, sample room toy with original tag, marked 1936, 4 3/8" long.

Left: *Coupe*, sample room toy with original tag, marked 1936, 4" long. Right: *Sedan* (two window), sample room toy with original tag, marked 1936, 4" long. These are economy models, a la Tootsietoy and other companies.

Sample room toys with original tags: *Hook & Ladder*, marked 1938, 8" long; *Fire Engine*, marked 1936, 4.75" long. Small toys for small budgets.

Coupe (take-apart), sample room toy with original tags marked 1936, 7.25" long.

Sample room take-apart toys with original tags marked 1936. Left: *Sedan with Farm Trailer*, 13" long. Right: *Sedan with House Trailer*, 13.75" long. *Courtesy of Hardin County Historical Museums, Inc.*

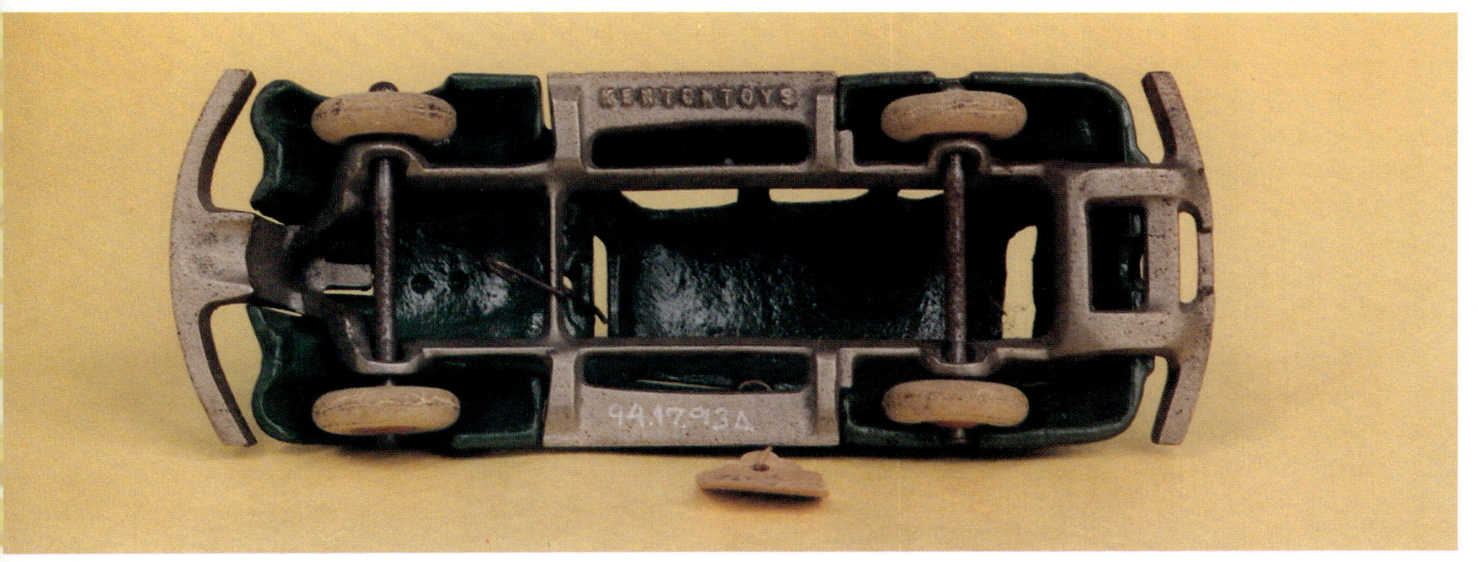

Bottom of *Sedan* showing take-apart clip mechanism which allowed body to be detached from chassis. *Courtesy of Hardin County Historical Museums, Inc.*

Sedan (Chrysler Airflow), sample room toy with original tag marked 1934, 5 7/8". The purple enamel was a dramatic departure from previous Kenton color schemes. The experiment began and ended with this automobile line. *Courtesy of Hardin County Historical Museums, Inc.*

World's Fair Bus, sample room toy, 14.5" long. "An exact reproduction, in snappy colors, of the busses now in use at The Century of Progress Exposition," boasted the 1933 catalog supplement. One of Kenton's most intricate castings. *Courtesy of Hardin County Historical Museums, Inc.*

Sedan (Hupmobile), circa 1935, 7.25".

Century of Progress Train, sample room toy with original tag marked 1934, overall length 22". This toy has been attributed to the Arcade Manufacturing Company, but recent discoveries suggest that it is indeed a Kenton product.

City Bus, sample room toy with original tag marked 1936, 7 3/ 8" long.

Outside the Kenton factory, Gene Autry and company vice president Willard Bixler, the man responsible for what was perhaps the best-selling cap pistol of all time, the *Gene Autry Repeating Cap Pistol. Courtesy of Hardin County Historical Museums, Inc.*

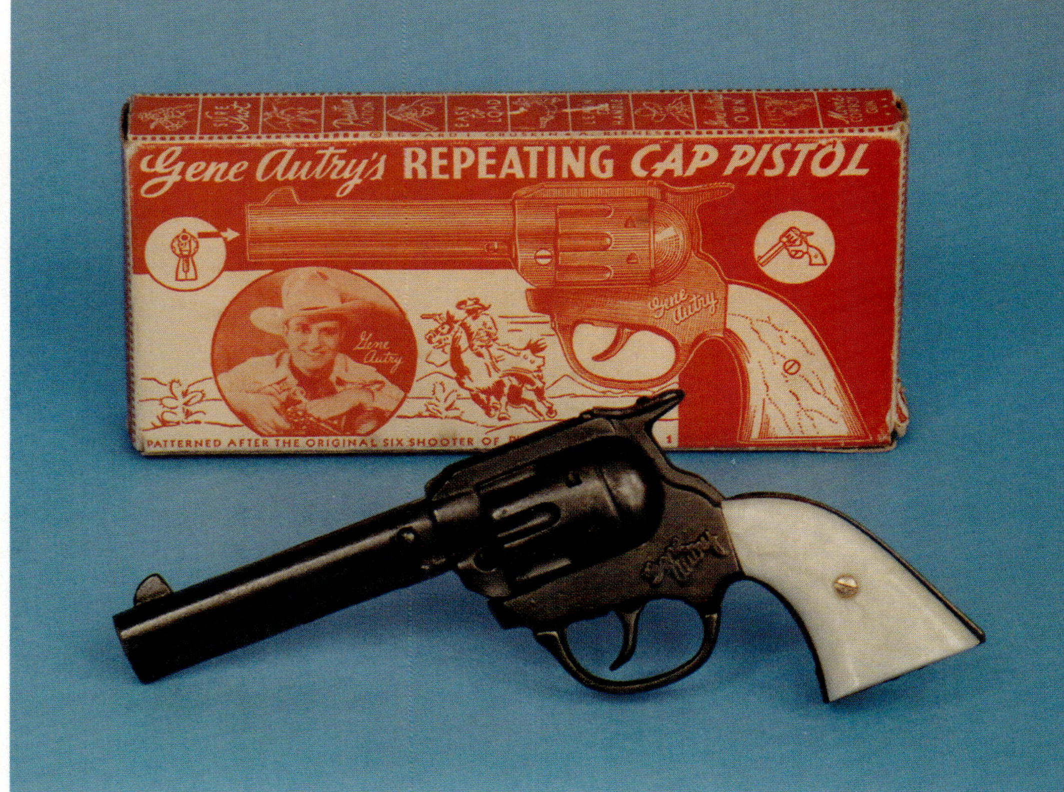

"The Sensation of the Decade" with original box; gunmetal finish, 7" long.

Selection of *Gene Autry Repeating Cap Pistols*, all sample room toys: (clockwise from upper right) *Junior*, gunmetal finish, 7.25" long; *Senior*, nickel-plated, 9" long; *Senior*, gunmetal finish, 9" long; *Senior*, nickel-plated, 9" long; *Senior*, copper finish, 9" long; *Junior*, gunmetal finish, 7.25" long. The red plastic grips are genuine Bakelite. *Courtesy of Hardin County Historical Museums, Inc.*

Gene Autry Junior, sample room toy, nickel-plated. The grip has a recess for a jewel, which was apparently never fitted. *Courtesy of Hardin County Historical Museums, Inc.*

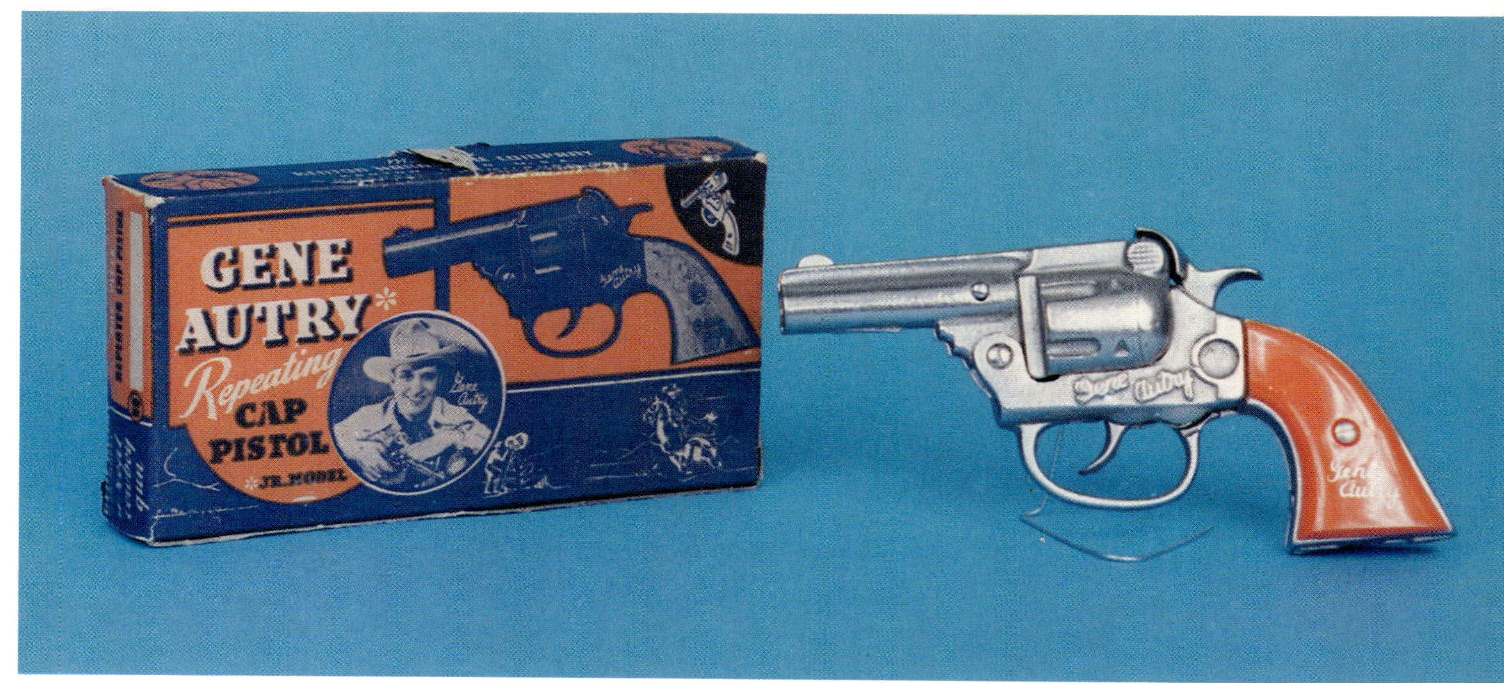

Display carton for the *Gene Autry Junior*, with its original contents.

Advertisement which appeared in Kenton newspapers the week before, and on, August 8, 1938.

Unique lobby card from Kenton Theatre, August 8, 1938. Signed by Gene Autry, Frankie Marvin, manager Gene Goodale, the Cass County Boys, and others.

Bandit Trail Game, 1940, 19" x 10".

Gene Autry with master pattern maker Joe Solomon during the plant tour on August 8, 1938. Note tools and toys, among them the *Rabbit Cart* and *Steeple Chase*, behind Solomon. *Courtesy of Hardin County Historical Museums, Inc.*

Man and women attach *Gene Autry* grips with screwdrivers. *Courtesy of Hardin County Historical Museums, Inc.*

Autry and assembler Monette Whitmore. *Courtesy of Hardin County Historical Museums, Inc.*

Company employees with Autry outside the plant on August 8, 1938. Lewis Bixler is second from left, with son Willard behind him. Harold "Buddy" Spencer, designer of the anti-aircraft machine gun, is to the right of the standing boy. *Courtesy of Hardin County Historical Museums, Inc.*

Trigger assembly. Chester Wachalec, the superintendent's son and successor, is seen behind Autry. Joe Solomon is at upper left. Also note riveter, upper right. *Courtesy of Hardin County Historical Museums, Inc.*

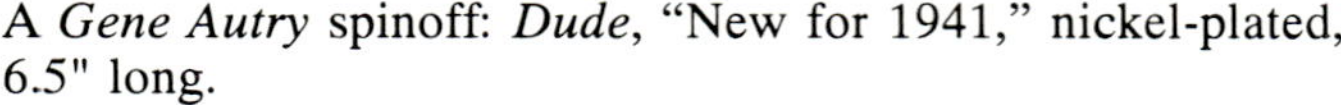

A *Gene Autry* spinoff: *Dude*, "New for 1941," nickel-plated, 6.5" long.

Circus Calliope, circa 1941, 14 1/8" long. An intriguing variation, with apparently original orange-capped driver and horses with a distinct gray hue. Metal front wheels are typical of the pre-war "Horse Drawn Vehicle of Yesterday."

The Horse Drawn Vehicle of Yesterday
Has Joined
The Streamlined Equipment of Today

IN OUR

DISTINCTIVE
AND
UNUSUAL
IRON TOYS

THE KENTON HARDWARE CO.
200 FIFTH AVENUE ROOM 548 NEW YORK CITY
FACTORY AT KENTON, OHIO

A 1941 advertisement for the "nostalgia line."

Circus Calliope, sample room toy, 14 1/8" long. Another version of the *Circus Calliope* with front rubber wheels. *Courtesy of Hardin County Historical Museums, Inc.*

Concrete Mixer, nickel-plated, circa 1941, 13 5/8" long. Yet another adaptation of the Jaeger prototype.

Combination banks. Left: *National* (with star) *Safe*, 3.5" x 2 5/8". Right: *Columbus* (with star) *Safe*, 4.75" x 3.75". The *National* and *Columbus* designs represented Kenton's last hurrah in banks. Painted versions which appeared in the 1941 catalog marked the end of the line.

In the late 1930s, the Warren Company of New York (a toy soldier manufacturer) adapted and sold Kenton *Roadster*s as the *Scout Car* (left) and *Staff Car*, both 7.25". The venture was short-lived; customizing ceased before the U.S. declared war on the Axis powers.

Warren *Scout Car* and the Kenton *Roadster* (sample room toy with original tag marked 1936) upon which it was based.

Willard Bixler, seeking "suitable war work," toured a traveling exhibition of defense industry products aboard a train stopped at Columbus. The only cast iron products to be seen were grenades. He was later surprised to find that no grenade contracts were available. Pictured here are possible experimental grenade prototypes or reference models. At left, a dummy acquired from a Kenton Hardware employee; the turquoise paint suggests a color used by the toymaker in the 1930s. At right, a casing found at the plant after it was closed.

"A Tough Way to Make a Living"

As a youngster, *Kenton News and Republican* carrier Clifford Keel didn't hesitate when the newspaper's "Rambling Reporter" gave him the opportunity to meet one of his matinee heroes.

"Carl Drumm–he was a nice, nice man–called me in his office one day with some other young fella–I can't tell you the other boy's name–and asked if we'd like to go to the Kenton Theatre and interview Gene Autry."

Recalled Keel, now 72, "I said, 'Oh, boy, I would!'"

When Gene Autry came to Kenton, Ohio, in 1938, Keel was one of the lucky youths who got to meet the cowboy star.

On August 8, Keel went to Gene Autry's sold-out show in the Kenton Theatre with reporter Drumm and a fellow top carrier. "After the show was over, we went backstage and I got to ask a couple or three questions. Mr. Drumm gave me one or two, and I had a couple in the back of my mind," Keel recalled.

One of the questions Keel asked the cowboy star was if he ever got hurt when he jumped onto his horse, Champion, to ride to the rescue.

Was Keel ever surprised when the Singing Cowboy displayed to the trio some of his battle scars. "There were some good bruises on the back of his legs, I'll tell you," said Keel. "I thought, that's a tough way to make a living."

Keel was a big fan of Autry's and enjoyed watching all the cowboy stars. "We used to go to the Kenton Theatre or the Ohio Theater on Saturday afternoons. We'd go to whichever one was playing Gene Autry, Tom Mix, Hoot Gibson, Tim McCoy, Ken Maynard. It used to cost us a nickel," said Keel. After the show was over, he and his friends would go home and play cowboys and Indians.

In spite of the fact that Keel dearly loved Gene Autry, he did not own a *Gene Autry Cap Pistol* during the latter years of the Depression. "We used to go down to the scrap pile at the Kenton Hardware Company where his pistol was made," he recalled. "We'd find this half of a gun and that half of a gun, and we'd put them together and somewhat have a gun, even though there was no trigger and no fancy handle. We'd stick it in our pocket and pull it out when we were playing cowboys."

Meeting his idol was exciting to Keel. "I felt as though it was quite an honor," he said.

"What struck me is he was a very nice-looking man, average height with a nice build on him," said Keel.

And Keel never watched the cowboy star's movies in quite the same way. "I often worried about him when he'd start climbing on his horse and taking off like you-know-what," he said.

"He would chase 'em down. I don't know how he caught 'em, but he did."

Former newspaper carrier Clifford Keel, who interviewed Gene Autry in 1938, stands before the printing press at the *Kenton Times. Courtesy of the Kenton Times.*

Chapter 10
Swan Song

"METAL TOYS ARE HERE AGAIN," proclaimed *Popular Science* in an April 1946 headline. Across the nation, writers not only expected a massive surge in postwar toy sales, but an explosion in toy technology as well. By December, magazines were filled with photographs and descriptions of miniature mechanical marvels, including a calliope outfitted with electronic circuitry and radio tubes, a racer with a ram-air-intake engine capable of speeds exceeding one hundred miles per hour, and an even faster *Shooting Star* jet.

Parents with more traditional tastes and more limited budgets were assured that the many old-line companies with their many traditional toy lines would return. "Many of Santa's old favorites are back in toyland this year," observed Ann Usher in the December 1946 *Better Homes and Gardens*. The restrictions of the War Production Board and the price-fixing of the Office of Price Administration, which had resulted in four previous "cardboard Christmases" of costly and ersatz toys, were relaxed. This reopened the market to cast iron playthings.

However, the Kenton Hardware Company found itself all but alone in a once-crowded industrial arena. The cast iron toy production of the Dent (Fullerton, Pennsylvania) and Champion Hardware (Geneva, Ohio) companies did not survive the Depression. The Kilgore Manufacturing Company of Westerville, Ohio, and its "Toys That Last" didn't outlast the forties. Arcade, a survivor of the Depression and World War II, was sold to Rockwell Manufacturing Company in 1946, bringing down the curtain on half a century of excellence in toy design. The Hubley Manufacturing Company, makers of bomb fuses during the war, abandoned the relatively expensive and sometimes elusive cast iron.

Competition from Stanley

The Stanley Toy Company of Oconto, Wisconsin, was one of the few manufacturers to offer Kenton serious competition in the shrinking cast iron toy market of the postwar era. In 1946 Stanley unveiled "A Really New Idea in IRON TOYS . . . authentic scale reproductions of horse-drawn vehicles out of the Gay '90s." This new idea was really nothing new, as Kenton toymakers could argue. However, there was no argument that it was successful: not long after the introduction of the Stanley surreys and fire department, the Wisconsin firm would boast it was the "LARGEST MANUFACTURER OF METAL HORSES AND WAGON TOYS."

No More Modern Ideas?

In the dawn's early light of the Atomic Age, Kenton must have looked like something of a living fossil. Kenton retained a logo which incorporated the *Graf Zeppelin*, monoplane, and other vehicles that were anachronisms to the new age. It stopped producing altogether any toys that had the semblance of appealing to boys and girls with "modern ideas."

While Kenton stuck with cast iron, Hubley and other major economical toymakers had turned to plastic and diecast zinc alloys. Inexpensive lithographed tin enjoyed a renaissance. The combination of cheap materials and cheap labor would bring prosperity to the reconstruction economies of Germany and Japan.

Kenton not only faced new technologies and new competition, but continued metal shortages as well. Still banks, long a Kenton staple, were dropped. Cap pistols were reintroduced in 1946, and carried the company up to 1949, when the factory was finally able to resume a full line of toy production. As late as 1951–the middle of the Korean War–Henry Katz, Inc., the Kenton sales office in New York, advised jobbers that all orders were subject to "availability of materials, restrictions, and government allocations." An addendum to a catalog of that same year indicated that all "single shots [pistols] were withdrawn Jan. 30 because of nickel restrictions."

Gene Autry to the Rescue . . . Again

In the summer of 1946, licensing agent Mitchell J. Hamilburg advertised, "Gene Autry, America's Number One Cowboy, is 'Back in the Saddle Again.'" The Singing Cowboy returned with his cap pistols. Lewis Bixler, L.S. Bixler's son, recalled that in 1946 and 1947 much of the company's time and energy was devoted to the production of the *Gene Autry Cap Pistol.* He added that 1.7 million were sold during this period. In the late '40s, the savior and staple of the Kenton line continued to be advertised as "the Cap Pistol with the outstanding reputation. Just ask any youngster, they all know this model." Still outfitted with red and imitation pearl grips, the *Senior* model was packaged in the display box that had been basically unchanged since the pistol made its debut in 1938.

It is likely that as the 1940s came to an end, the appeal of the *Repeater* began to wear thin. Years and years of minor changes did little to freshen the toy's appearance. Atomic disintegrators, sparkling submachine guns, new-fangled plastic water pistols, and flashy gilt diecast revolvers endorsed by Roy Rogers and other rising Western heroes had made serious inroads into the *Gene Autry Repeater*'s market. A change seemed in order, if not overdue.

The *New Engraved Gene Autry Repeating Cap Pistol* graced the cover of the 1950 Kenton cap pistols catalog. Although the text, basically unchanged from the previous editions, still insisted "youngsters all know this model," jobbers had to know differently. These *Junior* and *Senior* repeaters had a new look more than a little reminiscent of contemporary diecast designs: floridly embossed and recessed, nickeled and gun-metaled cast iron, with grips featuring spurs, stars, and ten-gallon hats, all in relief. Just as Gene Autry had evolved from the matinee's satined and sequined personification of the American Dream (as described by a Gene Autry souvenir program of the day), the *Gene Autry Repeater Cap Pistol* had been transformed from a Plain Jane "plowhandle" to a buckish buntline.

If Kenton's managers had pinned their hopes on the *New Engraved* models, their optimism was short-lived. The innovative *Gene Autry Repeater* was dropped unceremoniously after one year of production, possibly in response to slow sales. The Leslie-Henry Company of Mount Vernon, New York, for years a licensee of Gene Autry holster sets and cowboy suits, acquired the exclusive franchise to manufacture Gene Autry pistols in 1951. Leslie-Henry quickly developed an "entirely new and different" diecast design.

Kenton had ridden the Gene Autry wave for more than ten years, and tried to adjust and make do without their old mainstay. The *Bulls Eye* and *Lawmaker* cap pistols, reworked from *New Engraved* designs, were offered as substitutes. But without a character tie-in or a gimmick, they were doomed from their introduction.

Lewis Sharps Bixler Dies

In 1951, the toy company lost another mainstay with the death of Lewis Sharps Bixler on January 8. Bixler had navigated Kenton past economic adversity to the top of the cast iron toy industry. After son Willard had succeeded him as president, L.S. had spent more and more retirement time in Florida and Michigan. However, he had never been able to detach himself completely from the toy trade. In the postwar period he had remained a guiding force for the Kenton Hardware Company.

Mrs. Willard Bixler recalled many years later that the death of Lewis Sharps Bixler had signaled the beginning of the end for the toy company. "My husband's eyesight deteriorated and eventually he went to the office accompanied by his seeing eye dog, Pal. . . . Work became progressively difficult–the family wearied of the business," she added.

Kenton had to press on in the postwar years without the services of two other members of the "old guard," following the deaths of plant superintendent Mike Wachalec (succeeded by his son Chester, or Chet) and master pattern-maker Joe Solomon.

New Input

In 1947, the Kenton Hardware Company had announced the appointment of Henry Katz, Inc. of 200 Fifth Ave., New York as its sales representative. Katz, the former promoter of Hafner Mechanical Trains and Buddy "L" steel toys, had demonstrated a Midas touch in the toy market. He had a gift for promotion, and brought some color and flash to the hitherto staid Kenton advertising and literature.

Katz saw promise and potential for growth in the nostalgic concept, and traveled to the Kenton plant, where he went through the pattern room ("morgue") and resurrected old–but not outdated–designs. As a result of his visit, Katz contended, Kenton turned out the *Covered Wagon* ("An exceptional value–it is actually two toys in one. When the cover is removed the toy becomes a dray wagon"), *Hansom Cab* ("Take an old fashioned spin in this beautiful reproduction . . .), *Sulky* ("This sulky will create sales as fast as its prototype . . .), and *Surrey with Fringed Top* ("You can convert this Sunday Surrey quickly and easily to a weekday Buckboard . . ."). Customers were reminded that they were all hand-decorated, at a time when the vast majority of American toys were being mass-produced and becoming more and more impersonal.

In 1951, the suggested retail price for a *Covered Wagon* was $4.49, and slightly higher for customers west of the Rocky Mountains. The *Circus Band Wagon* topped all the other Kentontoys at $6.95. In spite of the claim that Kentontoys were moderately priced, the prices of the cast iron toys did not favorably compare with those of toys in other materials. For about a dollar less than it cost to own a *Circus Band Wagon*, a child could enjoy an exciting 145-piece plastic and sheet metal *U.S. Army Training Center Playset* from Marx.

More Pressures

Even higher freight rates and production costs were taking their toll on the cast iron toy trade. Conversion to cheaper, lighter weight, die-cast production was considered, attempted on an experimental basis, and abandoned. Lewis Bixler, the son of president L.S. Bixler, observed, "Cast iron toys became priced out of the market."

The prices would go even higher if the toy company employees unionized, Kenton owners and managers predicted. Talk of collective bargaining, strikes, boycotts, quotas, and improved working conditions must have struck fear into the hearts of the men who had weathered the problems of the modern toy industry to keep the cast iron giant afloat.

External pressures were felt as well. The German and Japanese toy industries bounced back in dramatic fashion. Imports of inexpensive plastic and tin playthings escalated. The domestic trade looked to survive by adopting more efficient marketing systems and improved production techniques, as well as expanding old plants and building new ones.

To compound the problems for Kenton, shortages of raw materials continued to plague production. Spokesmen for the Toy Manufacturers of the United States of America projected an output of metal toys for 1952 at seventy-five percent of the 1950 level.

Not only were the raw materials in short supply, but the boxcars needed for their shipment were tied up. Toymakers everywhere reported bottlenecks in deliveries in and out of their factories. The National Production Authority advised that it would limit use of metals as well as rail traffic for the duration of the Korean police action.

Persuader nickel and gunmetal finish, 7" long. "Large size break type repeating pistol with revolving cylinder, easy to load and to operate. A heavy model that looks and feels like a real gun," according to a 1937 cap pistol catalog. Local tradition has it that the *Persuader* and its successor, the *Gene Autry Repeater*, were used in a number of Chicago area holdups.

1952-The Epitaph

It seemed that Kenton had a knack for postponing the inevitable. When the end finally did come, it was upon the heels of one of the company's most elaborate and upbeat publications, the 1952 catalog. Its slogan, "QUALITY LEADERS FOR 50 YEARS," seems in retrospect to be an epitaph. The closure of the company was accompanied by little fanfare.

In February 1952, Henry Katz reported that Kentontoys were "selling in very good volume." Within months, however, cuts in the last catalog were being rubber-stamped "CANCELLED" (apparently by the same stamp that had been put to use in 1907). In domino fashion, casting, assembling, painting, and shipping departments were shut down, inventory was liquidated, and the remaining sample room collection was sold to Ken Idle of Chicago. The office remained open for another year or so to finish up business. The Kenton Hardware Company was sold to Frederick B. Hill, Jr., of Columbus, Ohio, and his associates. The firm continued to exist on paper until November 7, 1956, when its directors and officers, by unanimous consent of all shareholders, elected to "wind up and dissolve" the corporation.

Not many years after the demise of the factory, Kenton newspaperman Jim Stephey visited the home of the last factory superintendent, Chester Wachalec, and his wife, Frances. Stephey was amazed and enchanted by what he saw–a collection of hundreds of Kenton cast iron vehicles, cap pistols, and novelties. Around him came to life snarling bears in circus cages, charging horses with flying manes, their drivers urging them on to greater speeds, brilliantly dressed band wagon players tooting as they rolled along. For him, these were the toys of "them thar days" as well as his own time. Their quality cast iron construction had rendered them "practically indestructible." And their whimsy and charm, he added, would go on forever.

Kenton Play Guns (Dummy Style)

Play Gun Pistols do not shoot caps. Since there is a tremendous demand in territories where regular Cap Pistols are prohibited by law, we have developed a complete line of Play Guns (dummy style). These Pistols are identical in every respect to the regular Pistols, except that the mechanism is so designed that it is impossible to explode or fire a Cap in the gun. The hammer and trigger operate and "click" the same as the regular gun and have all the "Play" value. There are big possibilities for the sale of these pistols in the above mentioned territories.

Play Guns (dummy style) are available for the following Pistols:

No. 22	**Kido**	**No. 55C**	**Gene Autry Jr.**
No. 24	**Western**	**No. 60**	**Gene Autry Sr.**
No. 55	**Gene Autry Jr.**	**No. 60C**	**Gene Autry Sr.**

Play Guns (dummy style) are not available for the following Pistols:

No. 35 Jr. Police Chief **No. 66 Jr. Police Chief**

When ordering, simply show letter "D" after regular pistol number. For example No. 55D would refer to No. 55 Gene Autry Jr., in the Play Gun (dummy style).

Prices for Play Guns are the same as regular Pistols.

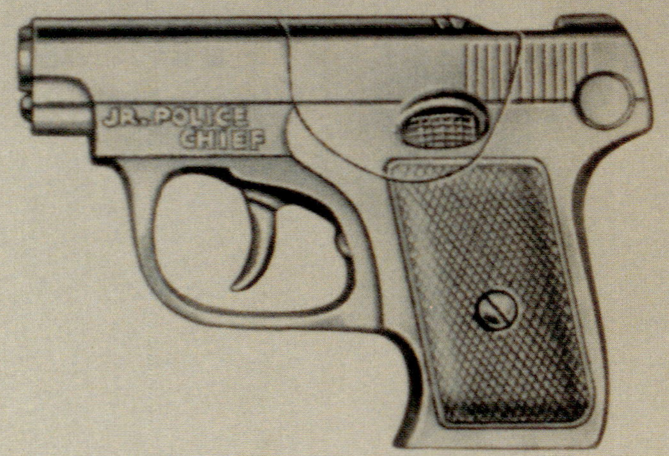

NO. 35 . . .
Jr. POLICE CHIEF

A very attractive model. Automatic type repeater with smooth easy trigger action. An exceptional grip feel that just fits the kiddy's hand. This is the best repeater possible in the low-priced field. Nickel plated finish only, 4½" long. Each pistol is packed in a two-color display carton with a cutout Police Badge and Secret Kenton Kode.

This is a great favorite with all children. One gross per carton. Carton weight 80 lbs.

Not available in Play Gun style.

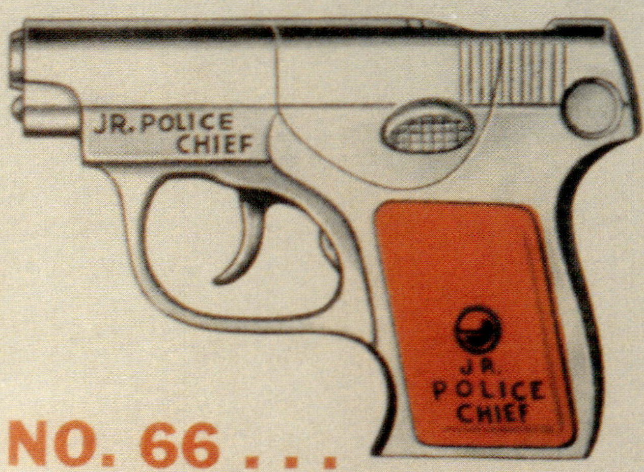

NO. 66 . . .
Jr. POLICE CHIEF

A new and improved pistol of the automatic repeater type with Red plastic grips as illustrated. Smooth and easy action. A perfect grip-feel for any youngster. An excellent value in low-priced pistols. Nickel plated finish only. Each plastic grip branded with the name "Jr. Police Chief". 4½" long. Each pistol is packed in a Silver and Blue two-color display carton with a cutout Police Badge and Secret Kenton Kode. One gross per carton. Carton weight 80 lbs.

Not available in Play Gun style.

KENTON'S
POPULAR
Gene Autry
REPEATER CAP PISTOLS
GENE AUTRY Sr.
NO. 60
The Cap Pistol with the outstanding reputation. Just ask any youngster. They all know this model. It is an exact scale model of Gene Autry's own six shooter. Furnished in nickel plated finish wth molded brilliant Red or imitation Pearl plastic grips, whichever is available. Gene Autry's script signature branded on each grip. 9" long. Each pistol packed in a three-color display box, showing a large photograph of Gene Autry. Packed 1/4 gross per shipping carton. Carton weight 37 lbs.
No. 60-C Gene Autry Sr. Same as No. 60, except furnished in imitation gunmetal finish with molded plastic grips of imitation Pearl on which Gene Autry's script signature is branded in Red.
Gene Autry
CAP PISTOL
NO. 55
GENE AUTRY Jr.
The Junior model and mate of the regular Gene Autry gun in a lower price range. Furnished in nickel plated finish with plastic grips of brilliant Red or imitation Pearl, whichever is available. Each grip is branded with Gene Autry's script signature. This pistol has break action and plenty of eye appeal for any youngster. 7¼" long. Each packed in a three-color display box. Packed ½ gross per shipping carton. Carton weight 55 lbs.
No. 55-C Gene Autry Jr. Same description as above, except imitation gun metal finish, with plastic grips of imitation Pearl.

Police Chiefs and *Jr. Police Chiefs*, including sample room toys. Top, left to right: *Jrs*. in nickel, gunmetal, nickel and black painted grip, and nickel and red plastic grip finishes, all 14.5" long. Center: *Police Chief* and *Jr. Police Chief* boxes with cut-out police badge and "Kenton Secret Code." Bottom: *Police Chief* in all nickel finish, center, flanked by nickel with plastic grip finishes, all 4 5/8" long. The *Police Chief/Jr. Police Chief* line was born in the heyday of Elliot Ness, and died with the factory shutdown in 1952. *Courtesy of Hardin County Historical Museums, Inc.*

Previous double-page photograph:

Centerfold of Kenton cap pistols in 1949 catalog.

Opposite page:

The *Gene Autry New Engraved Repeating Cap Pistol*, as promoted by the 1950 catalog.

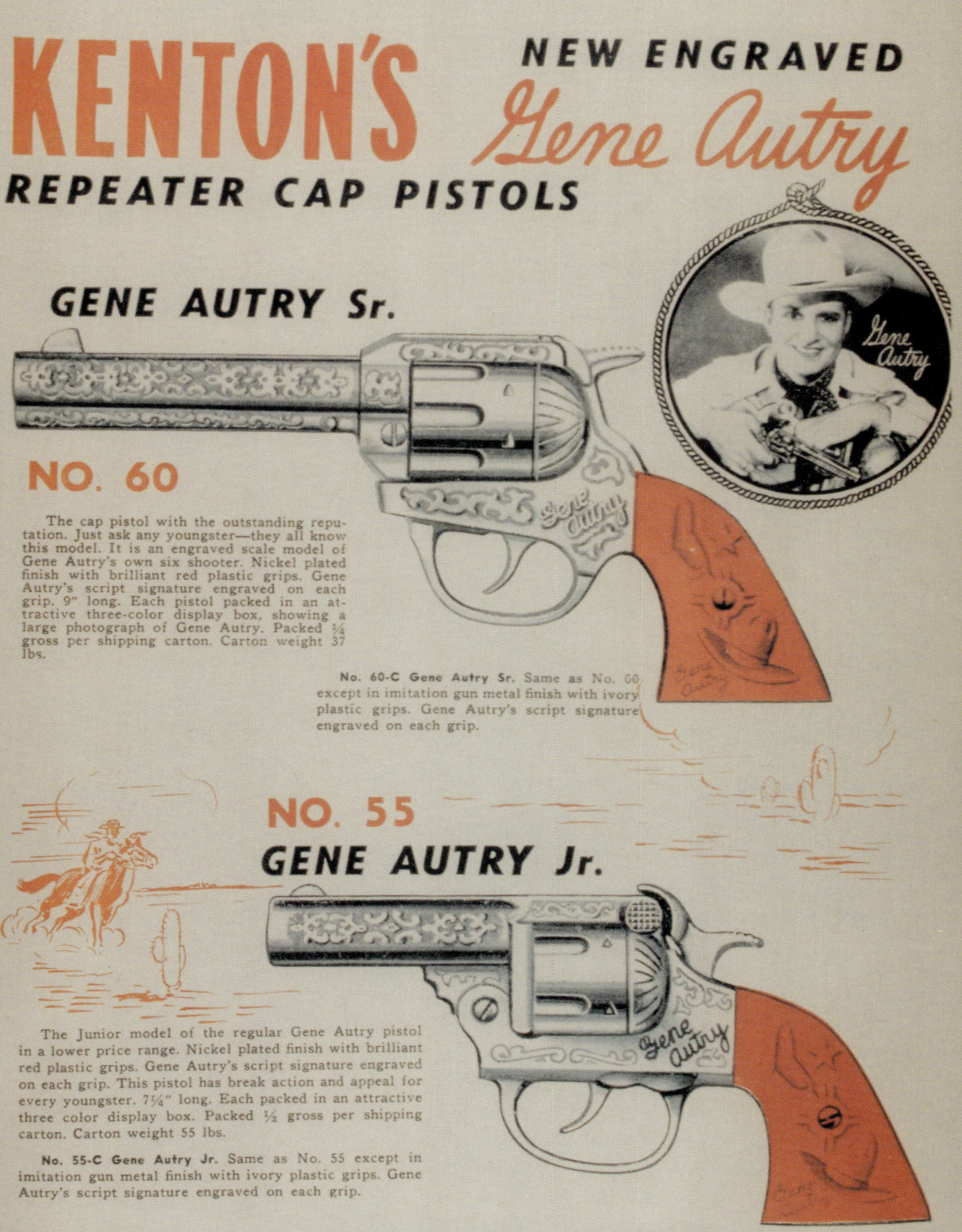
KENTON'S
NEW ENGRAVED
Gene Autry
REPEATER CAP PISTOLS
GENE AUTRY Sr.
Gene Autry
NO. 60
The cap pistol with the outstanding reputation. Just ask any youngster—they all know this model. It is an engraved scale model of Gene Autry's own six shooter. Nickel plated finish with brilliant red plastic grips. Gene Autry's script signature engraved on each grip. 9" long. Each pistol packed in an attractive three-color display box, showing a large photograph of Gene Autry. Packed 1/4 gross per shipping carton. Carton weight 37 lbs.
No. 60-C Gene Autry Sr. Same as No. 60 except in imitation gun metal finish with ivory plastic grips. Gene Autry's script signature engraved on each grip.
NO. 55
GENE AUTRY Jr.
The Junior model of the regular Gene Autry pistol in a lower price range. Nickel plated finish with brilliant red plastic grips. Gene Autry's script signature engraved on each grip. This pistol has break action and appeal for every youngster. 7 1/4" long. Each packed in an attractive three color display box. Packed 1/2 gross per shipping carton. Carton weight 55 lbs.
No. 55-C Gene Autry Jr. Same as No. 55 except in imitation gun metal finish with ivory plastic grips. Gene Autry's script signature engraved on each grip.

The *New Engraved Repeating Cap Pistols*. Left: *Junior*, nickel-plated with raised plastic grips, 7.25" long. Center: *Senior*, sample room toy, nickel with raised plastic grips, 9" long. Right: *Senior "Play Gun"* (non-cap firing dummy), nickel with raised plastic grips, 9" long. The new box incorporated a portrait of Autry that was at least a dozen years out of date.

Law Maker Repeating Cap Pistol, sample room toy, nickel with raised plastic grips. Note how the name "Law Maker" conforms nicely to the space formerly occupied by "Gene Autry."

Bulls Eye Repeating Cap Pistol, gunmetal finish with raised plastic grips, 7.25" long. Both box and gun, like the *Law Maker*, are *Gene Autry* makeovers.

Construction toys featured in the 1950 catalog. Left: *Road Roller*, sample room toy, 7.5" long. The actual road rollers were made in Galion, Ohio, approximately fifty miles east of Kenton. Right: *Road Scraper*, sample room toy with original tag, 6.75" long.

1950 CATALOG IN ACTUAL COLORS

KENTONTOYS are quality toys, moderately priced—an accomplishment from perfecting toy quality for more than a half century.

NO. 170 COVERED WAGON

A real Covered Wagon modeled from the famous Conestoga covered wagon which opened the West. An exceptional value—it is actually two toys in one. When the cover is removed the toy becomes a dray wagon. All young cowboys will demand this authentic toy.

Length 15 3/8". 1 pc. in box. 1 dozen per carton. Shipping weight per carton, 52 lbs.

THE KENTON HARDWARE COMPANY

FACTORY AT KENTON, OHIO, U.S.A.

SALES OFFICE • HENRY KATZ, INC., 200 FIFTH AVE., NEW YORK 10, N. Y.

The "two toys in one" *Covered Wagon* was the highlight of the 1950 toy line.

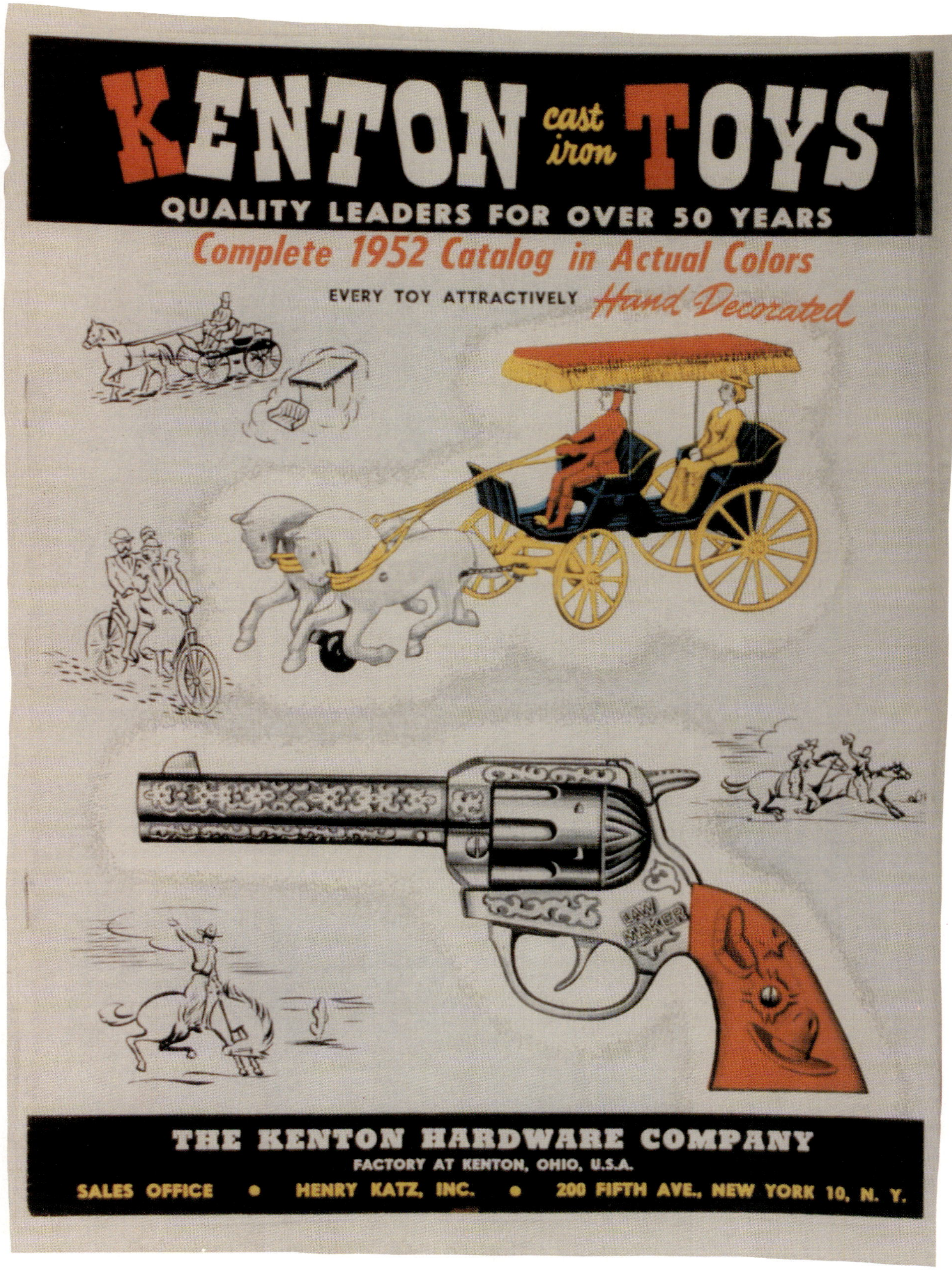

Last catalog issued by Kenton.

One-Horse Dump Wagon (Sand and Gravel), sample room toy, circa 1950, 13" long. *Courtesy of Hardin County Historical Museums, Inc.*

One-Horse Dump Wagon (Sand and Gravel), sample room toy, circa 1950, 15" long. *Courtesy of Hardin County Historical Museums, Inc.*

Surrey with Fringed Top, with cloth canopy and wire uprights, 1952, 13" long. *Courtesy of Hardin County Historical Museums, Inc.*

Milk Wagon, sample room toy, circa 1950, 12.75" long. "The counterpart of this toy is seen daily by thousands of boys and girls," Kenton catalogs maintained. How many milk wagons were still in service after the war?!

Sulky, sample room toy, nickel-plated wheels. A "nostalgia" item before and after the war. It gave up its nickel-plated wheels for painted ones during rationing. The *Sulky* was the first of the last: one of the first toys (if not the very first) to be deleted from the 1952 catalog.

Back: *Hook & Ladder*, sample room toy, circa 1950, 16.25" long. Front: *Hook & Ladder*, sample room toy with original tag, 11.5" long. Catalog cuts for these toys were enhanced by a line drawing of a building complex afire, eerily reminiscent of Kenton's own catastrophe in 1903. *Courtesy of Hardin County Historical Museums, Inc.*

Fire Engine, sample room toy, circa 1950, 13.25" long. *Courtesy of Hardin County Historical Museums, Inc.*

Hansom Cabs, sample room toy, 1952, 15.75" long. Apparently, the green variation is unique; some conjecture that it is a trial color prototype. *Courtesy of Hardin County Historical Museums, Inc.*

Two-Horse Stake Wagon, sample room toy, circa 1950, 10.5" long. *Courtesy of Hardin County Historical Museums, Inc.*

One-Horse Stake Wagon, sample room toy, circa 1950, 11.25" long. *Courtesy of Hardin County Historical Museums, Inc.*

Circus Band Wagon, lacks reins, circa 1950, 15.75" long. "Strike up the band! . . . The musicians, riders on the horses and driver, in their bright uniforms, are removable, increasing the play value of this toy," the 1950 catalog stated.

Circus Cage, circa 1950, 14" long. With the *Gene Autry Repeating Cap Pistol*, one of the toys most identified with Kenton. *Courtesy of Hardin County Historical Museums, Inc.*

Bear from *Circus Cage* with its probable celluloid toy inspirations.

Empty five-gallon paint buckets recovered from the factory by employee Harold "Buddy" Spencer. Gloss Black covered Kenton's last equine figures, while Tartar Red Light enhanced the fire departments and *Circus Wagon*s.

Pattern-Maker Recalls Factory Years

The Kenton Hardware Company's master pattern-maker, Joe Solomon, had enlisted the help of his apprentice, Richard Etherton, to design the official *Gene Autry Cap Pistol* in 1937. Etherton worked with Solomon, carving and molding soft metal and wax into a very close copy of the cowboy star's sidearm.

"The *Gene Autry* was made from scratch," Etherton recalled. "We had Gene's gun there in front of us."

But Etherton, now 80, remembered that he did not feel daunted by the task. "It didn't make us feel much different. I guess we didn't do as much thinking about it," he said.

Etherton said that Solomon was not optimistic about the prospects of the *Autry* toy pistol. "When we finished it up, Joe said if we sold twenty-five thousand of them we'd be lucky." The thought still brings a smile to Etherton's face. "I think I inspected about two million of them when I said to Joe, 'Remember that day when you said we'd sell twenty-five thousand?'"

The cap pistol became the top-grossing toy of the Kenton Hardware Company, carrying the company out of the Depression and through the difficult war and postwar years.

It took about a month to prepare the *Gene Autry Cap Pistol* patterns. Etherton remembers how he and Solomon took an iron rod of the proper size and bent soft white metal around it. "You build it up as you go. You have to carve certain parts and fill in behind them," he said. "After you get the outside the way you want it to look, then you have to start on the inside. That's when the fun starts."

According to Etherton, it was difficult to fashion the trigger so that it would trip the hammer to pull back. The *Gene Autry* pistol's automatic cap feed was also difficult to create. It was much easier to make a hammer and trigger for a single-shot cap pistol. "With single-shots, all you had to have is a hammer and trigger together," he said.

According to Etherton, making patterns for the Kenton Hardware Company afforded him a chance to invent. "I'm still working on one invention," he said. There is a mechanism in some toy pistols which causes the cylinder to revolve when the trigger is pulled. This mechanism is patented, and it is necessary to pay a royalty when it is used. Etherton said that he was convinced there was another way to accomplish the same end.

"I'm one of these nuts who figures there's got to be a way to do it," said Etherton. "I'm still working on a way to revolve the cylinder without the trigger as the primary mover."

Toward the end of the factory's existence in 1951, Etherton had turned his attention to another invention. "I'd been dreaming about automatic squirt guns with a water chamber in the handle. I'd been washing my hands, and I decided I was going to use a soap dispenser as a squirter."

He had just finished fabricating a squirter one day when L.S. Bixler walked out of his office holding an automatic squirt gun made by another manufacturer. "I thought, you dirty scoundrel," Etherton laughed. "I had an idea, but somebody beat me to it."

Etherton looks back fondly on his time at the Kenton Hardware Company, which spanned from 1935 to 1952, minus three years overseas during World War II. They were good years, he recalled.

"Joe [Solomon], well, he was a good fellow to work for," Etherton said. "He was strict about things, but he was good to work with. He was a good human being." Solomon and Etherton worked well together, with only occasional differences of opinion between the older and younger man.

Etherton remembers a lot about those years. "I can see that room where we used to work together–the window where he worked, the benches, the place where I worked. I can see the little tobacco can with chocolate in it for the little mice." Etherton got to the point where he could feed a family of mice which lived in the pattern room from his hand. "Joe thought I was nuts, but he didn't say anything," said Etherton. "We had a lot of fun."

Etherton said that Solomon let him learn by working on a *Radio Bank*, a *Milk Wagon*, a *Cement Mixer*, and numerous other patterns. "Those I was learning on. Getting educated or something," he said.

Recalled Etherton, "I had a job changing a colored man on a *Log Wagon* into a Dutchman, I think. Every time I carved on his face he ended up with chubby cheeks."

He said, "We'd take different ones like that and change them from year to year, change a little bit of something on them."

Etherton remembers that some toys were difficult to work on because of their size. "One fire engine we had was about that long," he said, holding his hands approximately a foot and a half apart. "It was a *Hook and Ladder*. I hated to put those things together because they broke so easy."

Solomon died during the last years of the factory's operation, leaving the pattern work to Etherton to complete. "When he knew he wasn't going to go back, he took me and said, 'Well, you do the best you can,'" Etherton remembered.

The first toy Etherton made all by himself was the *Surrey with Fringed Top*. "I was working with Joe before that," he said.

He figures he did most of the *Band Wagon*. "I suspect I made most of it because he had me make all the pieces to it and solder them together. He did part of the fancy work and I did part of it."

This 'fancy work' involved carving away at soft metal and building details with wire and wax. "If you wanted a figure, you had to find something and make it," Etherton said. For instance, the company's *Duck Bank* was patterned after a sugar bowl which belonged to Frances Wachalec, the wife of superintendent Chester Wachalec. The sugar bowl was used to make an impression in fine sand. The impression was then filled with the soft metal, which the pattern-makers could work on and refine.

Etherton blames the closure of the factory on the implementation of a union. The union negotiated for an agreement which provided that molders, who were paid by piece work, would lose no more than two percent of their pay when their work did not turn out.

"Scrap was running about sixty percent, and as far as the molders were concerned, they didn't care how they poured it. Two percent was all you were going to lose, even if you mispoured all of them, and the company was getting stuck with all that scrap."

Said Etherton, "It seemed to me there were some troublemakers in that union." He added, however, that he, too, joined the union because of the higher wages and the better chances for advancement.

During this time, most toy manufactories were getting away from cast iron to produce toys in lighter metals and in plastic. "I don't know what happened, because they were talking about putting in diecast machines," said Etherton.

Etherton remembers being directed to make a special toy pistol pattern. He was directed to make the pattern a certain thickness, and he knew from the measurements that the specifications were for a diecast pistol, although he was not told so.

"I made a pistol similar to the *Gene Autry*, and that's the last I saw of it. I don't know if they sold it, or what happened to the deal," Etherton said.

He maintains that the *Gene Autry Cap Pistol* was the last pistol the company ever made, although the catalog shows other pistol offerings. "They could have had somebody else's gun in their catalog," he said.

Etherton said that the company did not get enough outside jobs to keep production going. "They would get outside jobs from other companies making hog ringers and drawer pulls and things like that," he said. Ironically, it was with the manufacture of such products that the company had begun.

Etherton suspected that one day the toys he and his mentor made would be worth some money. "With me it would stand to reason after a certain length of time it would probably be worth something. I've always known what's one man's junk is another man's treasure."

Etherton refused to let projections about the toys' future value interfere with the work at hand, however. "At that time, I'd think about it, but I didn't let it carry on."

Richard Etherton was a pattern-maker for the Kenton Hardware Company. Today, at age 80, he uses his skills to create elaborate marquetry designs. The man who single-handedly made the pattern for Kenton's *Surrey with Fringed Top* stands beside an example of his more recent work. *Courtesy of the Kenton Times.*

West facade of the factory as it appeared in the fall of 1995. The structure serves as an ignition coil assembly plant owned by Andover, Inc.

Chapter 10 Epilogue— A Photographic Survey

The Kenton Plastics Corporation, owned by Howard Rome, attempted to capitalize on the Kenton Hardware Company's reputation by producing a line of plastic, battery-operated automobiles in the mid-1940s. *The Kenton*, 10" x 4", described as the aristocat of toy automobiles, featured a "Vibro-Roll" motor.

The Hardin County Archaeological and Historical Society (today the Hardin County Historical Museums, Inc.) contracted with Culp to produce an aluminum copy of the Kenton *One-Horse Stake Wagon* for the Hardin County Sesquicentennial Celebration of 1983.

Kenton Hardware products have been reproduced by several domestic and foreign toymakers since the late 1950s. In 1959, Frank L. Culp & Sons of Lexington, Ohio, sold a line of "authentic productions from original patterns." Sears sold reproductions of Kentontoys in their 1971 *Wish Book.*

The annual Gene Autry Days, sponsored by the Hardin County Chamber of Commerce, were initiated in 1994. The weekend celebration features displays of cast iron toys and Western-style entertainment.

Kenton's Country Connection Club and the Ohio Historical Society erected a historical marker in front of the old toy factory in 1991. The text was written by the author.

The Wachalec-Stevens Kenton Hardware Company Sample Room collection is a highlight of the Sullivan-Johnson Museum (flagship of the Hardin County Historical Museums, Inc.) in Kenton. The exhibit is comprised of more than two hundred cast iron toys and related products, a scale model of the factory, and a Wooton patent rolltop desk from the Kenton office.

Select Bibliography

Books

Aune, A.I. *Arcade Toys*. Brooklyn Park, Minn.: Robert F. Mannella, 1990.

Autry, Gene, with Michael Herskowitz. *Back in the Saddle Again*. New York: Doubleday, 1978.

Ayres, William S. *The Warner's Collector's Guide to American Toys*. New York: Warner Books, Inc., 1981.

Barenholtz, Bernard and Inez McClintock. *American Antique Toys, 1830–1900*. New York: Henry W. Abrams, Inc., 1980.

Davidson, Al. *Penny Lane: A History of Antique Mechanical Banks*. Oyster Bay, N.Y.: Al Davidson, 1988.

Duer, Don and Bettie Sommer. *The Architecture of Cast Iron Penny Banks*. Winter Park, Fla.: American Ltd. Productions, 1983.

Fraser, Antonia. *A History of Toys*. New York: Delacorte Press, 1966.

Friz, Richard. *The Official Identification and Price Guide to Collectible Toys*. New York: House of Collectibles, 1990.

Gottschalk, Lillian. *Toy Cars and Trucks*. New York: Abbeville Press, 1985.

History of Hardin County, Ohio. Chicago: Warner, Beers, & Co., 1883.

Harman, Kenny. *Comic Strip Toys*. Des Moines, Iowa: Wallace-Homestead Book Company, 1975.

Heide, Robert and John Gilman. *Cowboy Collectibles*. New York: Harper & Row, 1982.

Hertz, Louis. *The Handbook of Old American Toys*. Weathersfield, Conn.: Mark Haber & Co., 1950.

————. *The Toy Collector*. New York: Funk and Wagnells, 1969.

Ketchum, William. *Toys and Games*. Washington, D.C.: The Smithsonian Institution, 1981.

Logan, Samuel H. and Charles W. Best. *Cast Iron Toy Guns and Capshooters*. Davis, Calif.: "The Printer," 1990.

Long, Ernest and Ida and Jane Pitman. *Dictionary of Still Banks*. Mokelumne Hill, Calif.: Long's Americana, 1980.

Longest, David. *Toys, Antique and Collectible.* Paducah, Ky.: Collector Books, 1992.

McClintock, Inez and Marshall. *Toys in America*. Washington, D.C.: Public Affairs Press, 1961.

Moore, Andy and Susan. *The Penny Bank Book*. Atglen, Pa.: Schiffer Publishing Ltd., 1984.

O'Brien, Richard. *Collecting Toy Cars & Trucks No. 1*. Florence, Ala.: Books, Americana, 1993.

————. *Collecting Toys No. 6*. Florence, Ala.: Books, Americana, 1993.

Peirce, Bob and Shirley. *Iron Safe Banks*. Berlin, Wis., n.d.

Ralston, Rick. *Cast Iron Floor Trains*. Aiea, Hawaii: Ralston Publishing Co., 1994.

Rogers, Carole. *Penny Banks, A History and Handbook*. New York: E.P. Dutton, 1977.

Saylor, Robert. *The Kenton Hardware Company*. Still Bank Collectors Club, 1985.

Schroeder, Joseph J. Jr., ed. *The Wonderful World of Toys, Games & Dolls, 1860-1930*. Northfield, Ill.: Digest Books, 1971.

Whiting, Hubert B. *Old Iron Still Banks*. Vermont: Forward's Color Production, 1968.

Whitton, Blair. *Toys*. New York: Alfred A. Knopf, 1984.

Articles

Gurvis, Sandra. "Iron Toys: Gone But Not Forgotten." *Country Living* (December 1987): 6-8; 16.

Jacobs, Charles M. "Small Wonders, Big Business: Kenton's Cast Iron Toys." *Timeline* (July/August 1993): 18-29.

Long, Ernest and Ida. "Kenton Toys." *Collectors' Showcase* (November/December 1982): 39-43.

Newspapers

Ada Record, 1890-1900.

Hardin County Republican, 1901-1911.

Kenton Daily Democrat, 1900-1945.

Kenton Daily News, 1890-1891.

Kenton Democrat, 1890-1920.

Kenton Illustrated, 1890.

Kenton Press, 1898-1899; 1901-1904.

Kenton News–Republican / Kenton News and Republican, 1897-1952.
Kenton Times, 1954-1995.
Kenton Weekly Republican, 1893-1895.

Catalogs

Kenton Lock Manufacturing Company, 1892.
Kenton Hardware Manufacturing Company, 1900-1906, 1910-1912.
Kenton Hardware Company, 1913-1917, 1923-1933, 1936-1937, 1940-1941, 1949-1952.

Periodicals

Antique Toy World.
Penny Bank Post.
Playthings.
Toys and Novelties.

Unpublished sources

Kenton Hardware Manufacturing Company/Kenton Hardware Company Schedule of Selling Prices, 1907-1929, Sullivan-Johnson Museum Archives, Hardin County Historical Museums, Inc., Kenton, Ohio.
Wachalec-Stevens papers, Sullivan-Johnson Museum.
Kenton Hardware Company Articles of Incorporation, 1912. Secretary of State, Columbus, Ohio.
Kenton Hardware Company Certificate of Dissolution, 1956. Secretary of State, Columbus, Ohio.

Price Guide

Things are only worth
what one makes them worth.
– Moliere, *Les Precieuses Ridicules*

As a curator and an historian, I have found the task of compiling this price guide to be quite daunting. Throughout my career, I have focused on the cultural and educational value of objects. Traditionally, toys have been valued as "folk art"; more recently both curator and collector have come to regard them as valuable mirrors of our techno-industrial development.

Josiah Wedgwood, the great eighteenth-century ceramicist and author of *Dearness and Cheapness*, affirmed, "All works of taste must bear a price in proportion to the skill, taste, time, and expense attending their invention and manufacture." In assigning values to Kentontoys, skill, time, and expense were indeed taken into consideration, as were the ubiquitous supply and demand, topical appeal (comic, cowboy, and aeronautical themes, among others, have a broader appeal among collectors), geographics (e.g., Kentontoys, as one might expect, are very much in demand in the Kenton area), and condition.

The following prices (all in U.S. dollars) are for toys in **very fine** condition: not quite mint condition, yet not sullied by significant wear and use. Of course, well-worn, damaged or repaired/altered pieces should be valued accordingly. My own prices were guided by auction realizations, flea market and toy show offerings, and advertisements found in collectors' journals. They are not intended to be taken as gospel; many centuries ago, Publius Syrus observed that everything is worth what its purchaser will pay for it (and, might I add, what the vendor will sell it for).

Thus, the following are simply guidelines. The most important thing is to enjoy collecting and to love toys for their intrinsic value.

Page	Position	Value ($)
10	B	l: 250; r: 250
11	C	250
13	C	400
18	T	150
18	C	150
18	B	l: 20; r: 50
19	TL	100
19	TR	t: 50; c: 50; b: 50
19	BL	100
19	BR	75
20	C	300
21	B	25 each
23	B	150
24	T	150
24	B	1,200
25	TL	750
25	TR	3,200
25	B	3,000
26	TL	3,500
26	B	1,200
27	C	700
28	T	650
29	T	600
29	B	2,000
30	T	1,200
30	B	2,500
31	TL	400
31	TR	75
31	B	l: 150; r: 75
32	C	200
33	T	250
33	B	l: 100; r: 100
34	TL	50
34	TR	75
34	B	400
35	T	400
35	B	200
36	T	l: 3,000; c: 3,500; r: 3,000
36	B	tl: 3,500; tc: 3,000; b: 2,000
37	T	400
37	B	t: 400; bl: 400; br: 100
40	T	300
41	T	2,000
41	B	t: 2,000; c: 2,500; b: 1,800
49	B	3,500
50	T	3,000
50	BL	5,000
50	BR	t: (from left) 800; 1,200; 800; 1,500
50	BR	b: (from left) 800; 800; 750; 750
51	T	4,500
51	B	4,500
52	T	tl: 400; tr: 600; b: 500
52	B	l: 600; r: 600
53	C	3,000
54	T	400
54	C	400
54	BL	750
54	BR	150 with stand
55	TL	4,000
55	TR	1,800
55	B	l: 200 with stand; r: 100 with stand
56	L	1,800
57	L	no established price
57	R	no established price
58	T	100
58	B	250 (as standard toy)
59	T	5 each
59	B	100
62	L	250
63	L	250
63	R	150 (without grate)
64	L	100
64	R	125
65	L	150
65	R	no established price
66	TL	no established price
66	TR	200
66	B	450
67	T	200
68	TL	150
68	TR	250
68	B	(from left) 800, 400, 500
72	C	100
73	T	200
73	B	75
74	C	1,200
76	T	(from left) 125, 250, 150
76	B	350
77	T	30 each
77	B	l: 75; r: 100
78	T	2,500
79	B	l: 250; r: 250
83	T	10
86	BL	60
86	BR	75
87	B	100 with box
88	C	325
89	T	150
89	B	t: 400; c: 350; b: 350
90	C	325

91	T	l: 300; r: 200
91	B	200
92	C	200
93	TL	200
93	TR	200
93	B	300
94	T	50 each car
94	BL	1,500
94	BR	1,200
95	T	2,000
95	B	1,100
96	T	600
96	B	l: 650; r: 750
97	T	450
97	B	t: 450; b: 75
98	T	l: 500; r: 450
98	B	1,200
99	T	3,000
99	B	l: 300; r: 300
100	T	400
100	B	1,200
101	T	600
101	B	750
102	T	l: 2,000; r: 1,800
102	B	1,000
103	T	1,600
103	B	200
104	T	l: 1,000; r: 500
104	B	l: 700; r: 500
105	T	l: 2,400; r: 2,200
105	B	250
106	T	400
106	B	500
107	C	500
108	T	800 each
109	T	700
109	B	600
110	B	3,500
111	T	l: 1,200; r: 750
111	B	250
112	L	1,800
113	R	1,500
114	T	400
114	B	1,000
115	T	650
115	B	t: 500; bl: 400; br: 450
116	T	l: 450; r: 650
116	B	700
117	T	800
117	B	1,200
118	T	l: 450; r: 500
119	B	100
120	C	25-50
121	T	no established price
121	B	no established price
122	T	25
122	B	200
123		clockwise from top: 75, 50, 125, 160, 50
124	T	150
125	B	no established price
129	B	400
130	TL	no established price
130	TR	400
130	B	l: 40; r: 400
131	T	l: 350; r: 100
131	B	l: 250; r: 100
132	T	l: 350; r: 350
132	BL	l: 20; r: 20
133	TL	100
133	TR	300
133	B	l: 250; r: 300
134	T	l: 250; r: 200
134	B	1,200
135	T	l: 1,400; r: 1,500
136	T	400
136	B	1,600
137	C	700
138	T	1,200
138	B	750
139	T	250 with box
139	B	clockwise from top: 100, 175, 150, 150, 200, 100
140	T	175
140	B	250 with box
142	T	no established price
143	B	350
146	T	100
146	B	900
147	T	850
147	B	1,200
148	T	l: 75; r: 100
148	B	l: 2,000 with box; r: 2,500 with box
149	T	l: 2,000 with box; r: 2,500
149	B	no established price
153	B	85
156	T	t: from left, 60, 60, 60, 60; m: 25, 20; b: 75, 60, 75
157	T	200
157	B	150
158	T	l: 150; c: 300 with box; r: 200
158	B	200
159	T	200 with box
159	B	l: 450; r: 400
160	C	500
162	T	400
162	B	500
163	T	600
163	B	400
164	T	250
164	B	t: 400; b: 350
165	T	450
165	B	l: no established price; r: 500
166	T	350
166	B	300
167	T	1,000
167	B	750
168	T	l& r: 10; c: 25
172	T	25 each
172	B	60
173	C	200

Index

(**Bold face** indicates an illustration)